THE STORY OF THE LAMP
(AND THE CANDLE)

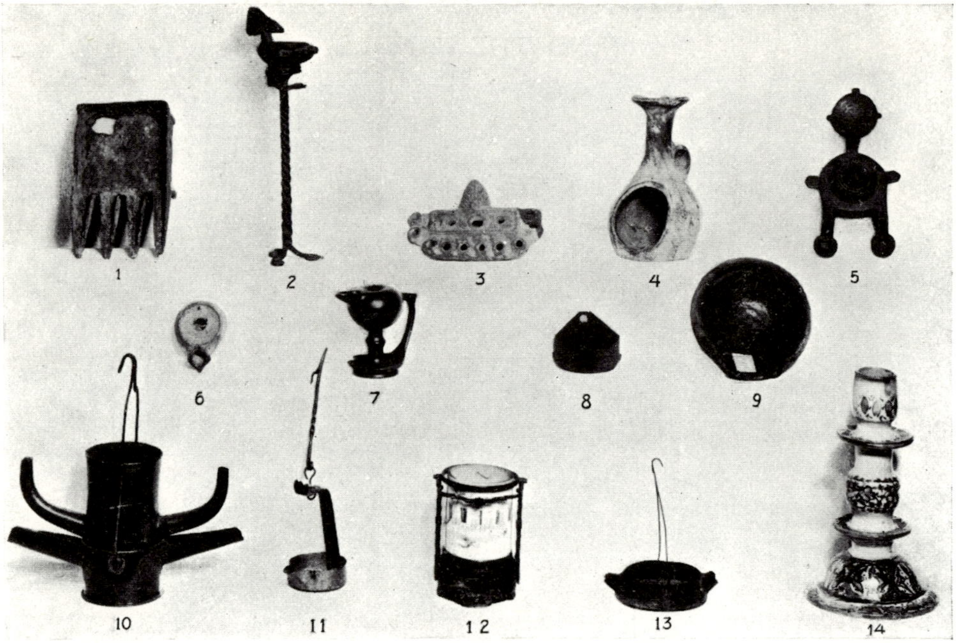

VARIOUS LIGHTING APPLIANCES

THE STORY OF THE LAMP (AND THE CANDLE)

BY

F. W. ROBINS

FELLOW OF THE ROYAL GEOGRAPHICAL SOCIETY,
MEMBER OF THE MUSEUMS ASSOCIATION,
EGYPT EXPLORATION SOCIETY, ETC.

With 50 illustrations,
showing nearly 500 specimens,
mostly from the author's
own collection

KINGSMEAD REPRINTS
BATH

First Published 1939
Oxford University Press
Reprinted 1970
Kingsmead Reprints
Rosewell House
Kingsmead Square, Bath

SBN 901571 33 4

Printed by lithography and made in Great Britain
at the Pitman Press, Bath.

PROLOGUE

MY excuse for tackling such a task as is involved in the compiling of a history of lighting appliances throughout the world is that the story of artificial lighting from the beginning (the end is not yet) is one of the few gaps in history which nobody else seems to have really attempted to fill. There are a few books on period phases and one or two abroad on national lighting history—none, so far as I know, on the whole field.

I do not claim that this book is a complete encyclopaedia of lighting appliances of all ages and countries. I have used every possible source available to me—public and private collections at home and abroad, including my own collection of some eight hundred specimens, references in books, excavation reports, papers read before learned societies and travellers' reminiscences, written and spoken. The most I hope for, however, is that this book will be a substantial nucleus to which additions may be made; to be quite sure of having covered the whole subject fully and accurately one would have had to travel in every quarter of the globe, no matter how remote, for which, unfortunately, life itself is too short, even if one had the resources.

Having regard to the intimate part artificial lighting plays and has played in family, social, and even ceremonial life, it is curious how little attention has been given to its history. Even contemporary references to the means of illumination are, in most ages and countries, extremely few and casual in their nature. Still more remarkable is the fact that painters and draughtsmen have constantly depicted night life in their times without showing the source of the light without which that night life could not exist. Nor do the artists of the Dutch school, with all their regard for detail in their interior scenes, remember to show the lighting appliances which must have stood in most of the rooms they depicted (save, in one or two exceptional instances, a brass candlestick).

For all this neglect of a very useful servant, the lamp and its relations have been an inevitable part of the equipment of every home practically ever since such a thing as a home, in the most primitive sense, has existed. It has played its part in prolonging the hours during which work has been possible and in that way, in many cases, removed delays from the march of progress. It has advanced the cause of learning; is not the phrase, 'burning the midnight oil', eloquent of that fact? It has played a most essential part in promoting and assisting social intercourse, as well as attended festivities in a decorative capacity. Lastly, but by no means leastly, despite his apparent indifference to its importance in word and picture, man has, perhaps unconsciously, recognized its importance in his life and its value as a gift of Providence by using it again and again as a symbol in his ceremonies and celebrations.

There is a world of philosophy to be extracted from the lamp's history if that were the purpose of this book, but to follow that line extensively would be to obscure the more material object of tracing history and development. One aspect, though, calls for special notice. It is the firm belief of the author that man's make-up and natural tendencies are, in general, very similar in all periods and parts, and very strong evidence of this is given by the way he goes to work when thrown back on primitive resources, whether he knows anything of what his predecessors did under similar circumstances or not. The lamp gave at least two examples of that during the Great War of 1914–18. One case occurred in the Ypres salient and possibly elsewhere as well; there, certain troops having been served out with sardines which proved to be unfit to eat (not perhaps an altogether unique experience), they made use of them another way by punching holes in the tins, inserting pieces of string as wicks, and turning them into fish oil lamps. It is extremely doubtful whether, in doing so, any of the men concerned realized that they were following precisely the same line of reasoning as some of their primitive remote forebears. In the other case, soldiers serving on the Gallipoli peninsula utilized condensed milk tins filled with bacon fat as lamps, thus producing, in modern days, the equivalent of one of the most ancient of all lamps—the open lamp with animal fat fuel.

A host of stories and sayings attest the way in which the lighting appliance is interwoven with the ordinary life of the people, from the bible story of the wise and foolish virgins to the expression, 'burning the candle at both ends'. Reference is made in Chapter XIX to the possible origin of the word 'booby' applied to a person. An Italian informed the writer that, in his native Calabria, there was a saying to indicate extreme poverty that a man 'used the oil from his lamp for cooking'.

It is rather startling to find that, apart from questions of detail in the design and construction of lighting appliances, there had been no important change in methods of illumination from the first invention of a primitive lamp until about a century and a half ago. Up to that time torches were in common use and such lamps as existed were simple reservoirs, in some cases even open bowls, containing animal fat or vegetable oils. Mineral oil may have been known and used in Mesopotamia at an early date but it was certainly not in widespread use, probably owing to difficulties of transport, apart from the fact that the only supplies would have been from seepages. In northern and western Europe, the common fuels towards the end of the eighteenth century were whale oil (sometimes seal or fish oil) and tallow—in the Mediterranean lands, from remote times to recent days, olive oil. The first real change came with the advent of gas lighting, which preceded both colza oil and paraffin, the latter indeed making its debut after the earliest experiments in electric lighting and not long before the introduction of that means of illumination in practical form! Yet the common impression to-day seems to be that paraffin lighting is something earlier in date than either gas or electric lighting!

That was one lesson the author learnt in his researches. Another was the curious filtration of Scandinavian or Teutonic culture through eastern and central Europe and, like a tide, lapping the northern slopes of the Pyrenees, Alps, and Balkans, sometimes even washing over them, while the European lands of the Mediterranean betray oriental influences in their lamps foreign to the north. Historic events have made their mark on peasant and rustic lamps—Arab influences in South Italy and Spain, Spanish influences in the Low Countries, Phoenician influences wherever the makers of those influences traded and colonized, Greek and Arab influences in Egypt, ideas from the Mahommedan north in West Africa, designs in native lamps from West Africa which betray the presence of Dutch traders there, and so on. History, geography, ethnology, science, art, the story of commerce and industry—all have their reflections in the story of the lamp. Naturally so, as it was almost as intimate a human possession as the dog.

F. W. R.

CONTENTS

CONTENTS

SECTION III. SIDELIGHTS

LIST (and details) OF PLATES

LIST OF PLATES

All the specimens illustrated are from the author's own collection except those on Plates II, III *a*, IV, VII *b*, VIII, X *a*, XVI *b*, XIX *b*, XX *b*, XXI *a*, XXIII *b*, and XXV, and Nos. 1 and 2 on Plate XV *b* and No. 11 on Plate XXVII *b*.

SECTION I

FROM THE TORCH TO THE CANDLE

CHAPTER I

THE FIRE

As the purpose of this book is to trace the history of lighting only, it would be out of place to devote much space to the question of the domestic fire as such, but something must be said about it, since it was certainly, at first, the only means of artificial illumination available. The discovery of the art of fire-making was not a single one. If any proof of this fact was needed, it would be offered by the several distinct methods employed to produce fire in different parts of the world, varying with the means available. Probably the most prevalent are the spark and tinder method (which would seem to have been the logical outcome of making flint implements) and friction methods, either between wood and stone or wood and wood. The use of mirrors and burning lenses, though less primitive, dates back at least to the fifth century B.C.

It is almost impossible to imagine any human progress having taken place at all without the primal discovery of the means of producing fire at will. Almost every human art and craft depends in some degree on the existence of a fire. Pottery-making, one of the oldest of human activities, was directly associated in the first instance with the domestic hearth. Metallurgy, of course, would have been impossible and there could have been none of the momentous discoveries of the nineteenth and twentieth centuries, including the use of electricity and the internal-combustion engine or any other modern mechanical contrivance. We should have continued to eat our food raw, as did early Palaeolithic man, and have acquired that food by the use of wood and stone implements. Our wars would have been largely matters of bludgeoning and stone-throwing, varied by short-range shooting with stone-tipped arrows and stabbing with pointed sticks. If we think too much of the last side of the question, however, we may be in danger of obscuring the far bigger balance of benefactions the knowledge of fire production has given us. One of these—though in some respects perhaps not absolutely an unmixed blessing—is artificial lighting and the consequent possibility of turning night, more or less, into day.

The domestic or tribal fire, in its stationary form, was at first the sole means of keeping at bay the deeper shadows of the night and has remained so among some savage nations. Incidentally, of course, it also served to keep at bay the more tangible forms of prowling animals, and it is a marked fact that, whereas fire and light have become friends and most useful servants to man, they still hold the brute creation in awe, which certainly seems to suggest that the ancients were not far wrong in regarding fire as a direct gift from the divine powers to man and a symbol of supernature.

The next stage would inevitably be the plucking of a burning brand from the fire to carry into the shadows, both as a means of lighting the way and as a protection. The first man who laid his hand to a piece of burning wood to pick it up was the originator of the torch, the development of which is traced in a subsequent chapter.

The difficulty of treating the subject of lighting in an absolutely chronological manner is not only the usual one of the various parts of the world failing to keep step with one another but also the survival, and even development, of primitive methods side by side with improvements in the same area. Such is the case with the fire, treated as a means of illumination. Long after torches, candles, and lamps had been evolved, fires were used for lighting, in braziers, as beacons, and in fire-baskets. In backward parts (the Dark Continent and parts of South America in particular) the fire and its direct offshoot the torch have remained in undisputed sway throughout the world's history.

According to Guhl and Köhner,[1] not only was light provided, in Homeric times, by fire-baskets or fire-basins standing on high poles and fed with dry logs of wood and splinters, but such fire-baskets on poles were still used by night travellers in South Russia and at night ceremonies in India at their time of writing (the first half of the nineteenth century A.D.). Neuburger, in *Technical Arts of the Ancients*, refers to the use, by the Greeks, of 'great pans of clay and copper, probably on plinths, as receptacles for very dry wood mixed with resinous wood'.

It might be thought that the advent of even so primitive a method of lighting as the torch would have rendered obsolete the fire, *en masse*, as a means of illumination, separately from its heating capacity. This was far from being the case, even in the more civilized areas of the world. Medieval Europe made considerable use of hanging fire-baskets and 'cressets'—using the latter word in its original application as denoting a form of fire-basket carried on a pole or standing on a pediment of some sort. These 'cressets' were employed by the 'watch' bands as late as the eighteenth century (*vide* illustrations from Hollar and in *Old England*) and survive in a stationary form (though more for heat than light) in the night watchman's 'fire-devil' of to-day. An adaptation of the cresset for stationary lighting consisted of a fire-basket at the top of a pole, with rungs on alternate sides for mounting by the attendants; this device appears in numbers of old illustrations and is incorporated in the punning arms of the medieval Bishop Beckington (Beacon-tun—a tun or barrel of fire used as a beacon) of Wells. This, too, was the medieval form of 'lighthouse', and even in the eighteenth century the mariners' light was still a fire in a stone or iron receptacle (an example of such a stone grate being still extant on the Scilly Isles).

Cressets, or fire-baskets on poles, have been used in Egypt, Arabia, Syria, Persia, and Morocco in the first half of the nineteenth century—in the Near East the fuel sometimes being bitumen. S. Burder, in *Oriental Customs* (1831), quotes Pitts as describing Arabian cressets, on poles, fed with short dry wood

[1] *Life of the Greeks and Romans.*

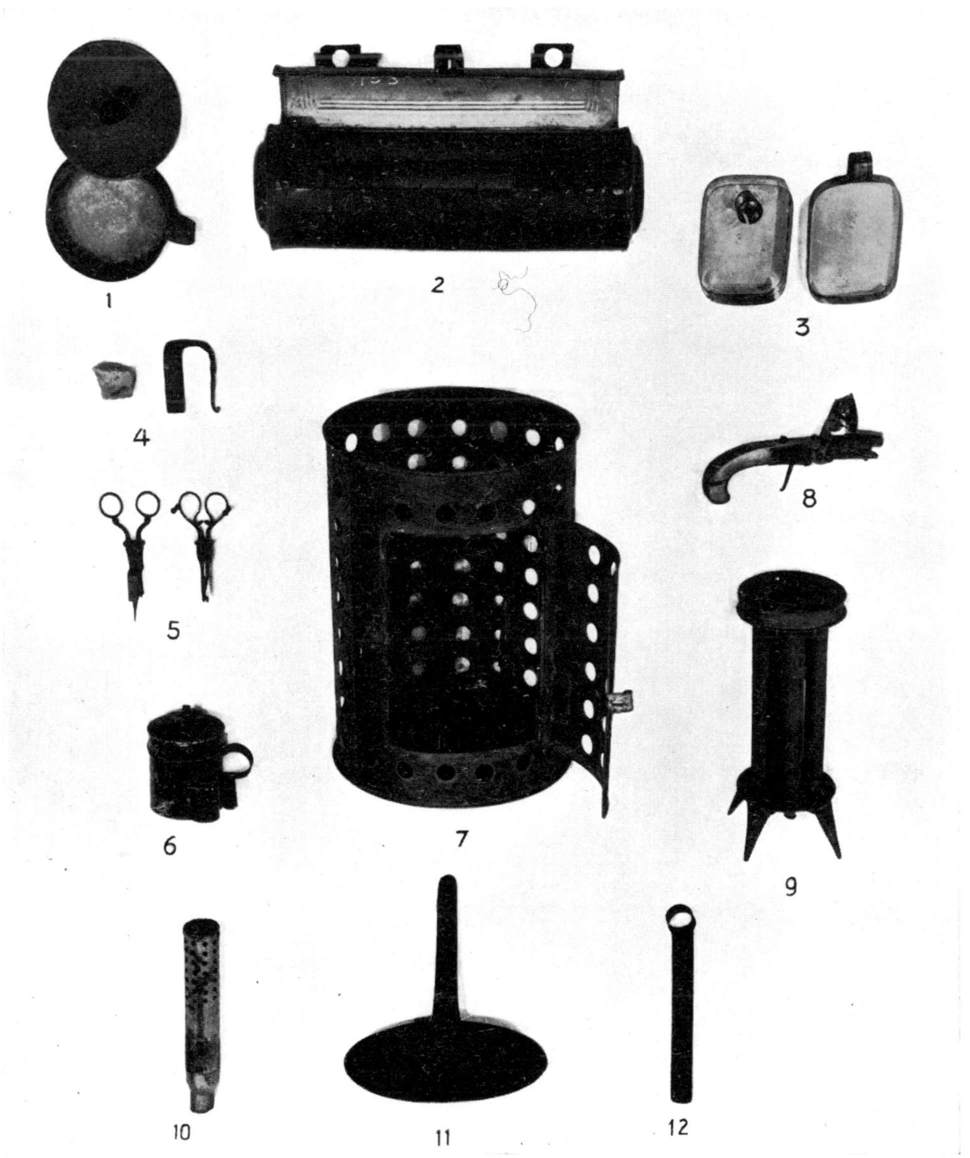

FIREMAKING AND OTHER ACCESSORIES

to light the way of the caravan and mark the resting-place of each traveller's section, each cresset, he says, being of a different form, to indicate the employer of the bearer.

In Bavaria, side by side with the pine-splinter holder akin to our rushlight holder, there was used, until comparatively recent times, an open lantern arrangement containing a small pile of pine splinters on a 'cradle' device, which was essentially a fire-basket. Into this category, too, comes the fire-basket torch of the early whaling industry, which survived until about 1880; this was an openwork basket of iron straps riveted together.

Braziers, in common use in all ages, while sometimes light-givers, have heating as their primary purpose and are rather outside the scope of this survey.

CHAPTER II

THE TORCH

IT would not have been very long after primitive man discovered the way to make a domestic fire when he found the way to create a portable light. A brand plucked from the fire to throw at a prowling beast, a child mischievously plucking a burning stick from the hearth, or a casual picking up of a piece of burning wood—any of these actions would have suggested the purposeful carrying of an offshoot from the communal fire into the darkness about, to keep the evil spirits away and give that extra courage that man, for all his present-day civilization, still feels when he has a light.

Herein lies a little problem. Which came first as a means of portable light, the thin splinter of resinous wood or the burning lump of timber forming a torch? The latter seems to be the more likely in prehistoric times; the split splinter would appear to have been a refinement and afterthought. On the other hand, the torch in archaic Greece seems to have begun as an aggregation of splinters bound together and, no doubt basing on this fact, Neuburger (*Technical Arts of the Ancients*) suggests that pine splinters preceded the torch as a means of illumination. He also suggests that the torch commenced its career as a firebrand used in war, instancing Assyrian sculptures of the ninth century B.C., but this is distinctly disputable; the torch was old by then and if its use as a source of light for the home and the way was preceded by anything it is more likely to have been the scaring away of beasts of prey. Primitive man would not have been slow to recognize the protection his fire gave him and the obvious fear of wild animals for the flames.

Lights of the torch class have an entirely different history from that of the lamp and their development is not into a lamp but into rushlights, tapers, and candles; in fact so imperceptibly does the torch merge into the taper or candle that it is sometimes difficult to draw the line of distinction. Broadly speaking, a torch may be said to be a wickless candle and a candle a torch with a wick, but even this is not a perfect distinction, since the link—an unmistakable torch—was an affair of treated rope (in other words, a heavy wick) while the vegetable 'candles' of South America are wickless.

One thing is certain: the torch was the first independent source of artificial light and antedated the most primitive form of lamp. This is evident not only in the history of lighting in classical lands but in the fact that even to-day there are backward races without lamps whose sole means of illumination are the fire and its offshoot, the burning brand. The scarcity of objects recognizable as lamps among stone-age deposits suggests that the torch was the commoner form of lighting right through the prehistoric eras. Outside the cave or hut it would certainly have been the wayfarer's companion if, in spite of his fears of mortal and spirit enemies, he had to venture abroad after dark.

It is curious that, despite the fact that lamps of stone, shell, and even pottery had been in use long before, archaic Greece seems to have had no lamps and depended mainly on torches for lighting. These were an advance on the brand plucked from the fire and were definitely manufactured articles consisting of sticks of resinous wood tied together with rushes, papyrus, or vine tendrils. They were treated with inflammable substances, such as resin, pitch, or wax, and, later, mixtures of such materials—including the resinous wood—were used as fillings to hollow cases of clay or metal, the torch so made being the Greek 'pharos'; one of these holders, in clay, was unearthed by Schliemann in the city of Tiryns, and ample evidence of the use of both forms of the torch is afforded by vase paintings, which also show the development of socketed torch-holders for the hand.

A little-noted fact is that the 'bound' form of torch used by the Greeks had a counterpart in a similar type of torch shown in the hands of Asiatic prisoners captured by Sennacherib and depicted on sculptures discovered by Layard.

In a country where wood was plentiful this naturally remained the principal basis of the torch for a long time. Necessity being the mother of invention, in a country where wood of a suitable nature was scarce some other material would be drawn into service sooner. There is still a lot to learn about the means of illumination employed by the ancient Egyptians, who possibly had lamps at an early stage. The 'new kingdom lamp' dealt with by N. de Garis Davies in the *Journal of Egyptian Archaeology*, April 1924, however, looks very like a torch of tallow or wax. It appears in two forms—one an almost diamond-shaped form and the other a cone or pyramid, but in both cases it is mounted on a staff for carrying or for standing. The pyramid form, it is suggested, stood on a flat base attached to the staff; except for the thickness and shape of the combustible material, it is very suggestive of the 'staff' torches of the Middle Ages. Generally, the torches are not shown alight, hence Mr. Davies suggests their use for ritual purposes alone. Side by side with them are to be seen 'candles' or 'tapers' of a twisted form, suggesting thick lengths of rope treated with oil or fat, such as were probably used in the 'candlesticks' (or more properly 'torch-holders') found in Tut-ankh-amen's tomb. Others, of tapering form, are strongly reminiscent of the dips of more modern times.

The isolated evidences of the use of torches or tapers in ancient Egypt do not suggest that they were anything like as common as in Greece. There they not only served the commoner purposes but figured largely in rite and ceremony. Fire has always been regarded by man as symbolic, both of divine providence and the home—naturally so, in view of its importance to human life and its place in the domestic circle. So, we find torches accompanying the Greek bride to her new home when the wedding evening procession conducts her to it, and her mother and mother-in-law starting the family hearth with them.

So also do they accompany the dead Hellene to his or her last home, and from this funeral custom of the ancient Greeks, carried on by their cultural heirs the Romans, is probably derived the taper accompaniment of the medieval funeral and the candles which burn around the Catholic bier to this day. (A similar idea of using torches in a bridal procession obtained in the East Indies, and there is a curious parallel to Greek practice in that the torches, of old rags soaked in oil, were in metal containers[1]—*vide* Burder, *Oriental Customs*.)

With the advent of the Greek lamp, about the sixth century B.C., the ordinary use of the torch naturally declined and its use by the Romans was mainly ceremonial. The inroads of northern barbarism into the territories of the Roman Empire brought it back into prominence again. The pinewoods of the north must have provided thousands of natural torches for the rude huts of Celt and Teuton alike. In what are known as the 'dark ages' the physical darkness of the night had the torch as almost its sole enemy and the halls of both Saxon and Norman were lit by it. (No wonder the 'red cock crew' so often on the roof-trees of the Saxon hall!) During the restoration of the wooden Saxon church at Greensted, Essex, in the middle of the nineteenth century, traces of lighting by torches were revealed, the timbers of which the nave walls were built showing scorching where lighted flares, moulded into lumps of clay,[2] had apparently been stuck on the walls. A flat hearth of burnt clay in the centre of the nave indicated that an open wood fire had also been used (*Discovery*, November 1937).

According to the Cluny Museum authorities, in medieval France a torch stuck on the floor or a chest lit even the man of comparative luxury to bed.

In the records of the Middle Ages the torch is almost inextricably mixed up with taper and candle; the name 'taper', now usually associated with a thin candle or wafer, was constantly used to denote what was really a waxen torch or large candle. (We are up against the difficulty of distinction between the one and the other here again.) For example, the torches shown at the feast in the British Museum illuminated manuscript Add. MS. 24098, f. 19b, appear to be tall candles, with wicks, set on poles, and these are probably the staff torches referred to in the Wax Chandlers' regulations. At the same time, however, there are regulations as to the offence of producing badly made links and these latter are probably torches pure and simple, with a wax basis.

In Cox's extracts from churchwardens' accounts there are given several examples of distinction drawn between candles or tapers and links or torches, as for example:

> 'Cost of buying wax and tallow and making of torches and candles together with links or torches, £3.4.10.' (Tavistock, 14th century.)
>
> '2 torches and 4 tapers, 10d.' (A funeral, St. Margaret, Westminster, fifteenth century.)

[1] See iron torch-holder from South India, Plate XVII (*b*). Cotton rags soaked in coco-nut oil were wound around the spiral support.

[2] Cf. the Cornish miner's method of fixing his candle!

PLATE II

(*a*) TORCHES BEFORE THE HINDENBURG CENOTAPH, MUNICH

(*b*) LINK EXTINGUISHER ON A BATH HOUSE

'2 torches of rosom.' (St. Edmund's, Salisbury, at Christmas.)

'A torch of rosin weighing 11 lbs.' (Provided by the Gild of the Chapel of Jesus Mass, St. Edmund, Salisbury.)

According to the same author (Cox, *Churchwardens' Accounts* in the *Antiquary* series), the torches referred to in these ecclesiastical reckonings were originally of twisted wax, but later the term was used for a coarse form of taper, largely mixed with resin, used in escorting a corpse to church and then to the graveyard, most churches keeping a stock and loaning them for funerals. Salzman, however, refers to resinous wood torches as used for ordinary purposes in the Middle Ages. Torches for important ceremonial occasions were, apparently, sometimes ornamental, as in the case of those used in the Corpus Christi fraternity's procession in medieval London—'100 torches of wax, "costly garnished".'

Leaving aside the confusion of terms in the Middle Ages, the torch proper, whether composed of vegetable matter (wood or rope) or with animal matter (wax or tallow), remained in use until long afterwards, even in civilized lands. The simplest form of all, the wood splinter, was in fairly common use in the seventeenth and eighteenth centuries in both Europe and the colonized North America. Scott, describing a Highland feast in *The Legend of Montrose*, says: 'Behind every seat stood a gigantic highlander, completely dressed and armed after the fashion of his country, holding in his right hand his drawn sword with the point turned downwards and in the left a blazing torch made of the bog pine. This wood, found in the morasses, is so full of turpentine that, when split and dried, it is frequently used in the highlands instead of candles.' 'Bog' candles of this type have also been used in Ireland, and the idea of human torch-holders was not uncommon at medieval feasts. A. H. Hayward (U.S.A.), in *Colonial Lighting*, states that the early New Englanders used pieces of resinous pitch pine, common in New England, cut in length and size like large candles, and stuck between the stones of the fireplaces or in improvised holders. These, he says, gave a bright flame but considerable smoke, were laid in stock for the winter, and were for a long period the common illuminant in the poorer homes of New England. He quotes Winter as saying, in 1642, 'Out of these pines is gotten the candlewood that is so much spoken of, which may serve as a shift among the poorer folk, but I cannot commend it for singular good because it droppeth a pitch form of substance where it stands.' An example of such a pine-splinter torch is described in the Smithsonian Institution catalogue as being of fat or heart pine (known in Virginia as 'lightwood') and burned in a pan stuck to the side of the kitchen fireplace. The catalogue in question states that such torches were to be seen at the time of publication (1928) used in the cabins of poor whites and negroes, and that a familiar sight in Louisiana 'in olden times' (no date given) was that of negroes carrying bundles of 'lightwood' for sale in New Orleans.

In Italy, for processional purposes in recent times, saplings have been taken and beaten into a fibrous state, fat being then rubbed into them to form torches somewhat akin to the Greek bundled type. These are probably survivals of the torches used by ancient Roman peasants (to which Virgil makes reference).

Turning to more localized forms of torches, the Smithsonian Institution exhibits a series of torches from various parts of the world which include one of the Mexican candle tree, a mass of resin attached to a rod used by the natives of Africa (in which construction there seems to be some echo of the Egyptian staff torches already mentioned), and a 'candle' from Kashmir made by winding a cord wick around the end of a piece of the stalk of some plant and forming over it a mass of tallow, the result presenting a form again rather analogous to the Egyptian torches, though in miniature form. (It is curious at how many points Egypt and India touch, in this question of lighting appliances in particular.) In the South Seas, the ever-useful coco-nut provides torch material.

Whole animals themselves have served as torches in different parts of the world. The fat bodies of stormy petrels have been used in the Shetlands (apparently a fertile centre for the production of primitive lighting ideas), with wicks thrust down their throats. The Greak Auk, now extinct, served occasionally as both fuel and light, and Joly, in *Man before Metals*, says the Danes of the kitchen midden period employed a wick of moss with one end buried in the fat-laden stomach of a great penguin. The candle-fish (north-west coast of America) was used by Indians; it was held in the cleft of a split stick (*vide* examples in the South Kensington Science Museum and Smithsonian Institution collections). The tail of the dogfish, cut into strips, was burned for light by fishermen off the banks of Newfoundland.

In South America, 'candles' (wickless and therefore more truly torches) are made of balsa wood charcoal mixed with wild beeswax, wrapped in leaves, the torch resulting being some 1½ to 2 inches thick and 2 feet or so long.

Both the American Indians and the East Indians have used torches formed from bundles of thin wood (pine in the west, bamboo in the east) which are strongly suggestive of the developed torch of ancient Greece. Bark and palm leaves are frequently used in torch manufacture. Again East meets West in the use of resin bound in palm leaves, a form of torch found in the West Indies, Africa, and the Far East. The Smithsonian Institution catalogues examples of this torch type from Singapore, Burma, Siam, the Philippines, Porto Rico, French Congo, and the Gaboon River. The use of torches of this class is widespread in night fishing, the light serving the purpose of attracting the fish as well as that of illumination and being carried either in the hand or in the bows of a vessel (usually, in the latter case, in a basket or cradle which links the torch with the 'fire-basket').

Obviously, the torch is an adaptable production whose constituents can be anything inflammable which is easy to procure in the locality concerned and one might therefore multiply recipes indefinitely. An English recipe was given by 'Torchbearer' in the *Daily Mail* during 1935 as 'fibre rope treated with Stockholm tar, costing one penny per foot, as used locally' (Warrington, Lancs.) 'in torchlight processions, with success'. The links or 'flambeaux' (the terms seem interchangeable) of Georgian days were on similar lines but with resin, tallow, and beeswax as the inflammable coatings, in combination or singly. Appropriately enough, Bath, city of Georgian houses and sedan chair associations, possesses a piece of one of the links in its Institution Museum and examples of link extinguishers on the front walls of several of its eighteenth-century houses (as well as some of later date and doubtful authenticity!). Links of the period, illustrated in *Old England*, however, are short, club-like, and smoky, suggesting the resinous wood type rather than the slender parallel flambeau of rope and wax shown in the Bath specimen.

[*See Plates I, II, III, and XVII.*]

CHAPTER III

SPLINTERS AND RUSHLIGHTS

SIDE by side with the torch carried in a socketed holder and the torch case filled with resinous material, there appears in the Greek vase paintings a third variety consisting of two pieces of wood fastened, in the form of a St. Andrew's cross, to a staff. This is an early form of splinter light.

A splinter of resinous wood lit at the end is, in effect, a thin torch; it is impossible to draw a hard-and-fast line and say that a piece of wood of one diameter is a torch and that of a lesser dimension a splinter; when, however, a piece of wood is systematically cut into strips by a tool or machine devised for the purpose, a refinement of manufacture has set in which justifies a difference in classification and when the strip so formed is given an inflammable coating the principle adopted allies it with the rushlight. Splinters of resinous wood, especially pine wood, carefully cut and dried and sometimes treated with combustible substances have been used for lighting from remote times. The form of the so-called candelabra of the Etruscans, with their spikes projecting from the tops of the standards, suggests the use of tapers or splinters impaled on them; the similar standards of the Romans, be it noted, are generally lamp-stands, as the spikes have been replaced in these by a flat table on which to stand a lamp, and the term 'candelabrum' applied to such a later device, derived though it is from the earlier one, is rather misleading.

In the forest areas of Scandinavia, Finland, Russia, Germany, and central Europe generally, as well as in Scotland, the use of pine splinters for illumination was common from an early period and remained in vogue until recent times. M. Ilin, in *Turning Night into Day*, writes of Russian peasantry: 'They would split up a straight-grained log without knots into slender pieces of kindling wood about a yard long and light them at one end. . . . This kindle light . . . continued in use for more than a thousand years. Not so very long ago it was still used in some out-of-the-way Russian villages.'

Originally, no doubt, the splinter light was hand-carried; an old Norwegian print shows a peasant farmer holding a utensil in either hand and a lighted splinter in his mouth, while his wife spins, also with a lighted splinter in her hand and a bundle of unlighted splinters tucked in her waist-belt.

A holder for the burning splinter was an obvious invention of necessity and probably took the form of a cleft stick in the first instance. The design of many of the splinter holders of north and central Europe strongly suggests this, apart from its natural logic. The crudest of these take the form of a wooden upright set in a wood block (or constructed from a branch whose forks provided three or four feet) for standing on the floor (*vide* examples in the Bergen, Moscow, and Budapest museums, for instance) and having at the top end a simple

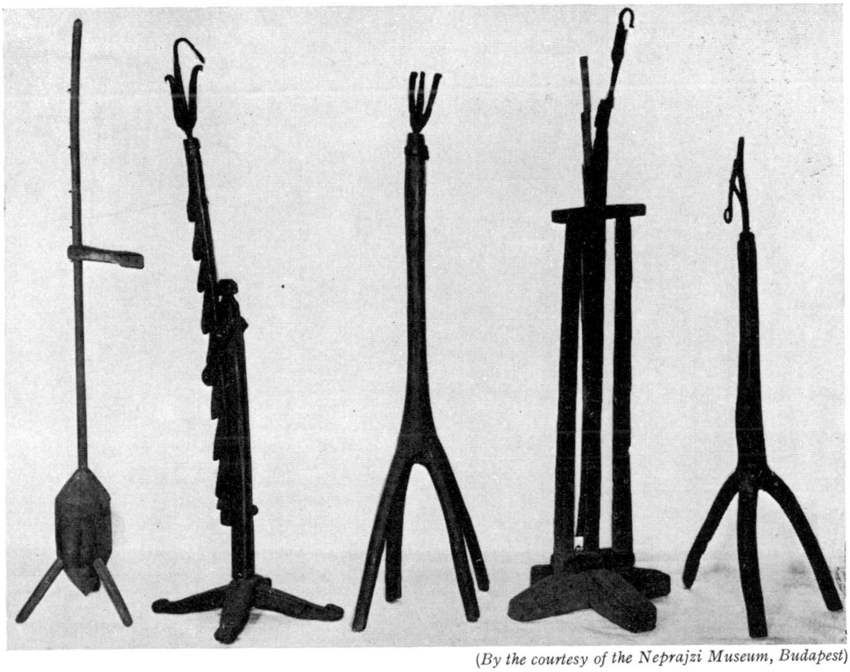

(*By the courtesy of the Neprajzi Museum, Budapest*)

(*a*) HUNGARIAN AND SLOVAK PINE SPLINTER HOLDERS

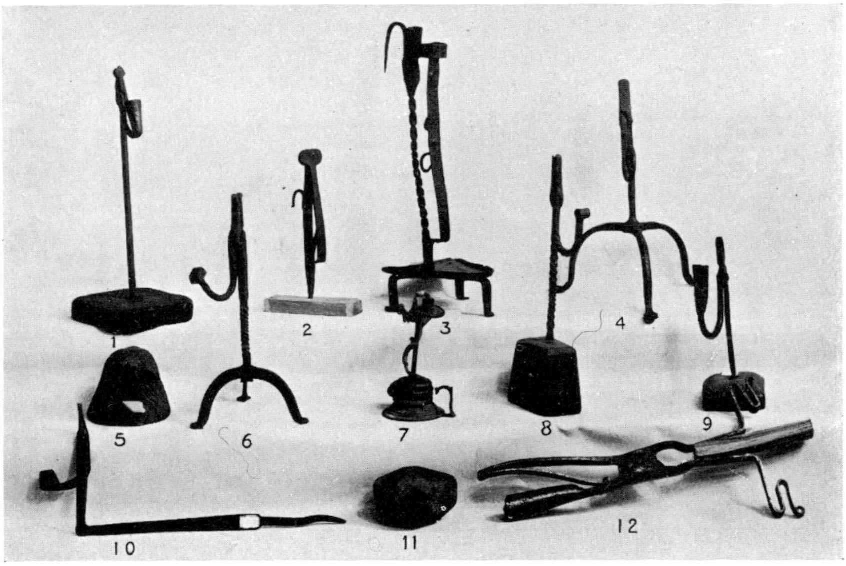

(*b*) RUSHLIGHT, DIP, AND SPLINTER HOLDERS

clamp or grip of two or three short diverging pieces of iron. Owing to the upward progress of a flame on a piece of wood the splinter had, naturally, to be held at an angle with the flame at the lower end; an iron plate was often placed on the floor beneath the holder as a precaution against fire. Ilin states that it was 'usually the job of one of the children to look after the kindle light while the grown-ups worked'.

Crude though the whole idea of the splinter light was, primitive science and the art of the craftsman got to work on the holders and both improvements and refinements set in. An early advance was the substitution of iron for wood in the body or shaft of the holder; the signs of burning on many of the wooden examples sufficiently attest the advantage of this. Iron standing specimens sometimes retained the wood block at the base but more often they terminated in feet or in an open circle of metal. On the art side, some of the tall splinter holders in the Moscow museum are fine examples of the blacksmith's art, with twisted stems and foliate heads; some of these have spikes rather reminiscent of the Etruscan candelabra.

A marked improvement in splinter holders was the provision of a clip or tongs at the top of the stand to hold the splinter, in place of merely wedging or jamming it between pieces of iron, or sometimes a loop was provided. At a late stage and in a few examples a ratchet arrangement allowed for the raising or lowering of the light, but by the time this had been added to the splinter holder the holder itself was merging into a country-side dip or candle holder, which often took a similar form, with the addition or substitution of a crude socket.

In the metal splinter holders a type appears for sticking into walls to form a bracket, instead of standing.

There is no trace of the use of splinter holders in England (no doubt owing to the widespread use of the rushlight and the scarcity of suitable resinous wood in the greater part of the country), but pine splinters or 'fir candles' were used as lights in Scotland, where the holder, in the form of an iron upright terminating at the top end with tongs and usually set in a wooden or stone base, was known as a 'puirman'—a title which suggests its own story of lowly economy. Sometimes a wooden stem intervened between the iron tongs and the base. The 'fir candles' were separated by a special knife or 'fir gully' and dried in a wooden or iron cradle known as a 'kelchin', suspended in the chimney. An alternative way of burning the pine splinters was in an iron 'chandler' or openwork cup of iron straps, forming a cradle, on a stand. (*Vide* examples of the 'puirman', fir gully, kelchin, and chandler in the Marischal College Anthropological Museum, Aberdeen University.)

The Scottish 'chandler' has its parallels in Germany and Scandinavia, where hanging trays filled with small pieces or chips of pinewood and surmounted by large hoods formed open 'lantern'-like appliances in rustic surroundings.

An interesting pine-splinter holder appears to be peculiar to Bavaria. This takes the form of a grotesque human head in clay, with the mouth open to receive the splinter. The Norwegian print already quoted at once leaps to the mind and proclaims the origin of the idea—the carrying of the splinters in the mouths of the peasants themselves (see No. 5, Plate III (*b*)). Another German version, in iron, has tongs for a splinter at one end and a socket at the other. Presumably this was a reversible arrangement for insertion in a wood block (No. 12, Plate III (*b*)).

The splinters or wafers used in all these devices were separated by knife, plane, or a special machine devised for the purpose and were generally of pine, boxwood, or birch, untreated. The floor stands seem commonly to run from 3 to 5 feet in height.

Closely allied to the torch and the candle and replacing the splinter in areas where resinous wood was not so plentiful is the rushlight. Its actual date of origin is obscure; it is reasonable to suppose that it antedated the candle, though the surviving utensils connected with its use are never earlier than the latter. It is a feature of the English and Welsh country-side, where the pine splinter was a stranger (since pines were not indigenous to southern Britain in historic times and only reached there, since prehistoric ages, within the last three centuries). It consisted of a peeled rush (properly, says Gilbert White, the *juncus effusus* or common soft rush) having one strip (or sometimes two strips) left for support, such strip, incidentally (and accidentally), having the added effect of the later plaited candle wick in curving the burnt material to one side and reducing the necessity for snuffing by bringing the carbonized point into contact with the air. The peeled rush was subsequently dipped in fat or grease to give it the combustible coating of a miniature torch or thin candle. Gilbert White gives a detailed account of the method. The rushes, having been gathered in late summer or early autumn, flung into water as soon as cut and peeled in the manner already mentioned, were dried in the sun before the dipping operation (assuming the sun deigned to shine of course!).

'A good rush which measured in length 2′ 4½″ burnt only three minutes short of an hour and a rush of still greater length has been known to burn one hour and a quarter. These rushes give a good clear light. . . . In a pound of dry rushes, we found upwards of 1,600 individuals. Now supposing that each burns only half an hour, then a poor man will purchase 800 hours of light, a time exceeding 33 days, for three shillings. According to this, each rush cost 1/11th of a farthing . . . but the very poor, who are always the worst economists, burn a halfpenny candle every evening, which, in their blowing open rooms, does not burn more than two hours. Thus, they have only two hours' light for their money instead of eleven.' (White, *History of Selborne*.)

What would the reverend gentleman have said to the present cost of candles? On this basis, if economy rather than candle-power is the desideratum, one might well supplant our gas mantles and electric light bulbs by the humble rushlight—always supposing, in these days of 'middlemen' and 'marketing', the said rushlight could still be purchased for one-eleventh of a farthing or thereabouts. White goes on to say that the rushes were gathered and prepared by decayed labourers, women, and children; the peeling, it may readily be understood, was a matter of acquired skill. The fats used were the leavings of the kitchen (particularly mutton or bacon fat) and White calculated that 6 lb. of fat would dip a pound of rushes or so.

In 'Farm Life in Old Sussex' (Maude Robinson, *Sussex County Magazine*, June 1931) a similar account of the method of making rushlights is given as obtaining in Sussex and an old woman is quoted as calling them by the apparently local name of 'fried straws'. Miss Robinson adds that a little beeswax added to the fat made the rush more cleanly to hold when carried by hand about the house and that her grandfather was in the habit of using a rushlight as combined pointer and light when conning his weekly newspaper.

When set in a holder, rushlights were held in a slanting position, which assisted the self-snuffing process; unlike the pine splinters, and for obvious reasons, they were lit at the upper end. The holder might be (as not long ago in the north of England and in Wales) merely a split stick (*vide* nineteenth-century specimens in the National Museum of Wales, Cardiff), but as early as the fifteenth century metal rushlight holders were in use. These, like the pine-splinter holders, were on iron feet or wood-block stands; the former, usually tripod, seem to be earlier in origin and connect with the form of some early candlesticks. The wood blocks range from rough flat squares to turned and polished stands. The grip end of the holder was in the form of a pair of hinged pincers, with a counterpoise weight on the free arm to ensure a firm hold. Later, the weight was provided by a socket to hold a dip (though the other form continued, so that the absence of the dip holder does not necessarily mean an earlier date). Two explanations of the addition are given—one that the socket was originally a dowser, subsequently adapted as a dip holder, the other that the dip being a more expensive illuminant, the combination provided for its occasional use while being available for the more constantly used humble relation, the rushlight. Possibly the first is the real origin and the second an adaptation of the idea to a different end.

Occasionally the rushlight or dip holder (particularly if continental) has a hook for holding utensils; this is notably the case with Breton examples in the St. Malo Museum.

[*See Plates I and III.*]

CHAPTER IV

DIPS AND CANDLES

THE true candle probably arrived in the days of the Romans. Egyptologists sometimes refer to 'candles' and 'candlesticks', but they would readily admit that their 'candles' are more correctly described as torches, though the possibility of a wick being present in some of the constructions of fat or wax must not be ruled out of account. The seven-branched candlestick of the Old Testament was most certainly not a candle-holder at all but a group of lamps, as is shown by the instructions for the provision of the olive oil for replenishing it.

In the Roman era a candle definitely appears, though at first pitch seems to have been used for its construction. According to Neuburger, Roman candles were of two forms, one resembling a torch, with a fibrous substance predominant, the wick twisted together of fibres of papyrus and other plants, and the other of modern form, having a wick of a species of papyrus and wax or tallow as combustible material; the wick was first impregnated with sulphur and then repeatedly dipped in liquid tallow or wax, the wax used being prepared with special care from honeycomb cleaned in water and dried for three days, afterwards being pressed out and boiled once more in the same water mixed with fresh cold water. Finally it was bleached by repeated boiling with sea-water and dried in the open air. Pliny, in the first century A.D., refers to Greek and Roman candles of flax threads coated with pitch and wax.

Despite the older ancestry of the candle, through the torch, and its own undoubted existence in Roman days, it does not seem to have been in such common use as the lamp for several centuries. Its great periods were the Middle Ages and, in this country at any rate, the sixteenth to eighteenth centuries. According to Sir Walter Besant, wax candles were used in the better-class Saxon houses—but they were probably luxuries. That the candle was a recognized illuminant in Saxon days is shown by King Alfred's ordinance: '. . . commanded his chaplain to supply wax in sufficient quantity and he caused it to be weighed in such a manner that when there was so much of it as would equal the weight of 72 pence he caused the chaplain to make six candles thereof, each of equal length, so that each candle might have twelve divisions marked across it. These, burned in succession, lasted for 24 hours and each division indicated one third of an hour.' It is said that Charles V of France had a time candle in his chapel to last the full 24 hours.

During the Middle Ages and right up to the early days of the nineteenth century, the commoner candles were made of tallow. In fact, the use of wax, in the form of beeswax, was confined almost entirely to the churches (and even in these for ceremonial and altar use only) until the cheapening of wax

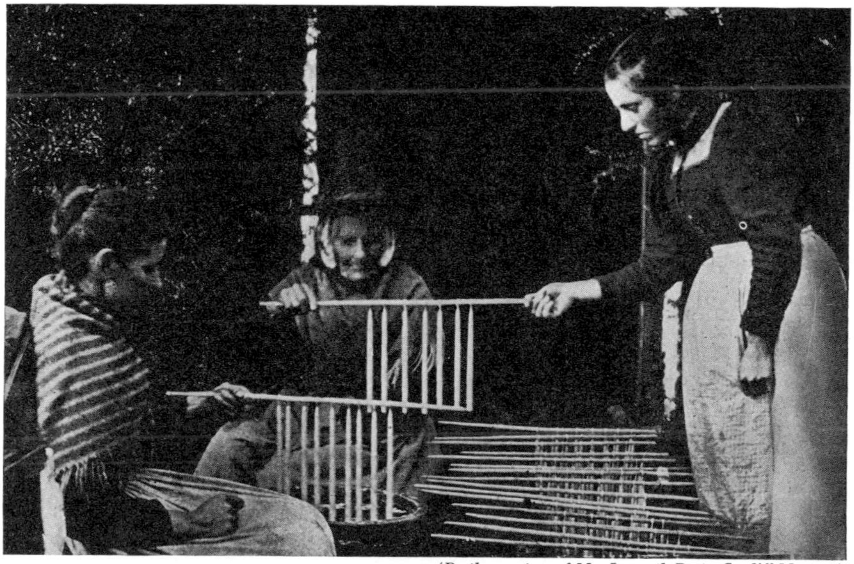

(*By the courtesy of Mr. Iorwerth Peate, Cardiff Museum*)

(*a*) WELSH WOMEN MAKING TALLOW DIPS

(*York Castle Museum*)

(*b*) A YORKSHIRE DIP FACTORY

in the fifteenth century increased its use. For the more exalted purposes, e.g. at the celebration of Mass, bleached beeswax was used, but at Masses for the Dead yellow wax might be employed (Cox).[1] The tallow candle was usually a home-made product and even the monasteries carefully saved the fats from their kitchens for the making of candles for ordinary use.

Waxcraft enumerates five different ways of making candles:

1. Dipping the wick and drawing it through melted wax or fat.
2. Building up by hand.
3. Pouring the melted wax on wicks. (This method was used for wax candles only, because of the tendency of wax to stick to moulds, and is still used for such beeswax candles as are now made for religious purposes.)
4. Pressing through a cylinder. (Drawing was used for small tapers—the wick being passed through fat, drawn through a hole, then dipped again and drawn through a larger hole.)
5. In moulds.

The first method was that most commonly used for tallow candles. An old inhabitant of the New Forest, however, told the writer of still another method employed by her father. This was to drive a bar into a bed of damp sand and pour the melted tallow into the hollow so formed, in which the wick was held suspended meanwhile.

The dipping method goes back to the Roman period and was the main method used during the Middle Ages and long after; it was frequently facilitated by the use of a frame from which a number of wicks were suspended, either in a row or otherwise, thus enabling several wicks to be dipped at once. Each layer of fat had to be allowed to set sufficiently before the succeeding dipping took place and an old Cornishman (now dead) used to tell how he and his numerous young brothers and sisters were set to dip candles before they could have their dinner; sometimes, in their hurry, they re-dipped the wicks too soon, with the result that the last layer was spoilt instead of a further layer being added.

Newburger says that sand was sometimes added to the tallow in the Middle Ages, to prevent excessive heating.

Wicks in home-made dips were commonly of rush but sometimes of textile materials and, where a frame was not used, rushes were usually tied together in bunches of four, so that one wick was held between each pair of fingers for dipping. After dipping, the candles were hung in a loft or cellar to harden and whiten. Mr. Iorwerth Peate, in the handbook to the Folk Collection in the National Museum of Wales at Cardiff, describes the making of 'water candles' (*cannwyll y dwfr*) in rural Wales:

> 'A number of wooden wands about 2 feet in length were whittled down to a thickness of $\frac{1}{4}''$ and 9 to 12 hanks of cotton twisted along each stick to form the wick. A loop was made around the stick and the cotton held out straight and plaited in two strands for about nine inches, the end being then sealed with a small piece of fat. A large cauldron with water in

[1] *Churchwardens' Accounts*.

the bottom was filled with fat and hung over the fire, the process being sometimes carried out in the open, upon a trivet. A ladder was supported horizontally on two casks or boxes. When the fat was thoroughly melted, the sticks were taken out one by one and the strands dipped in fat and hung from the ladder to harden. The process was repeated until the fat covered the strands thickly enough to serve as candles.'

The *Daily Mirror* of 30 November 1935 showed an illustration of a Swedish peasant woman making Christmas candles in similar fashion—even to the use of the ladder.

While in the monasteries of the Middle Ages, and no doubt in the great households as well, the staff manufactured their own candles to a considerable extent, occasionally an itinerant candle-maker would be called in to manufacture a stock. In Paris in the thirteenth century there was a guild of candle-makers who went from house to house to make tallow candles, the manufacture of wax candles being in the hands of a separate guild. (The French, incidentally, have different names for the two articles, the *bougie* being the wax candle and the *chandelle* the tallow kind.) With the formation of the guilds or companies of the city of London, too, the trades were recognized as separate. The Worshipful Company of Waxchandlers dates from 1358, the Charter of Incorporation being given in 1484, by which time the manufacture and use of wax candles had increased considerably, owing to the cheapening of wax; the Tallowchandlers had their charter in 1462. Adulterating the wax was an offence which was punished by the Company, the usual punishment being forfeiture of the wax, 'wrought or unwrought', a mild penalty compared with the 'punishments to fit the crime' inflicted by the guilds of other trades (such as the putting of a vendor of bad meat in the pillory, while the offending meat was burnt under his nose!).

The making and provision of candles looms large in the church accounts of the Middle Ages and the following centuries. There were the altar candlesticks (two large sticks flanking the crucifix and three smaller, two of which were kept burning during the service and the third during the most solemn part) to be filled with wax candles, candles to be burnt before images and in chantries, tapers or large candles for funerals, candles for special occasions, tallow candles for the ordinary illumination of church and domestic offices. At St. Edmund's, Salisbury, there were recorded, at the end of the fifteenth century, in the Lady Chapel to the south of the choir two altar lights and a lamp before the Blessed Virgin, sustained by the guild of wives; at the High Altar two lights and a lamp sustained by the guild of the Blessed Sacrament; at Christmas two torches of resom; over the Sepulchre at Easter, the great taper or Paschal taper, also 100 candles on prickets of beech; in the Chapel of Jesus Mass, two tapers for the altar and a torch of resom weighing 11 lb.—this fraternity also providing 6 lb. of tallow candles yearly for the Morrow Mass in winter; in the nave a specially supported Rood light, in addition to the Trendall or hanging corona of lights, as well as candles on the Rood beam. Each year special tapers were bought for carrying before the Blessed Sacrament at the times of visitation; at the close of the fifteenth century a lantern was purchased to shelter the tapers. At St. Mary-at-Hill in 1353 there were two tapers kept burning on the iron beam before the image of Our Lady at the high altar, on Sundays and holy days; two tapers before the Angels' Salutation of the image of Our Lady in the body of the church every evening; at the time of the singing of Salve Regina a fifth taper was to burn at the south altar between the figures of St. Thomas and St. Nicholas. (Cox, *Churchwardens' Accounts.*) Mr. Cox is probably correct in his statement that the provision of ceremonial and other lights was the most costly of all church charges in medieval England (or any other Christian country in the Middle Ages for that matter). The cost of these church lights was met largely by the funds provided by various guilds attached to the churches in towns. Sometimes there were endowments for the purpose; at Cowfold, in the fifteenth century, four cows belonging to the church sustained (financially, of course!) tapers for Our Lady, St. Katharine, and St. Anthony. Bequests were made for the maintenance of lights, fees were paid for the use of lights at funerals, and candle silver or wax silver was raised by means of collections made by medieval churchwardens in their parishes.

Candlemas Day (2 February), still marked in our calendars, was the day on which, from the eleventh century onwards, candles were blessed and the lessons to be drawn from them as symbols of burning love towards Christ were impressed upon the people.

There are several illustrations of candles of various lengths, carried in the hand, in the Luttrell Psalter—including scenes at the Presentation in the Temple and the death and burial of the Virgin Mary—and these are the items usually referred to as 'tapers' of course. They are plain candles of considerable length, but another illustration shows a shorter candle of a twisted form, held without a stick in a man's hand. Candles with moulded or applied decorations, painted or gilded, were used for special occasions at least as early as the Tudor era (*vide* specimen in the Tudor House Museum, Southampton). At this period Southampton had Town Chandlers (two divided the town between them in 1577) and price regulation—the latter evaded and a constant source of trouble. The current rate about that time seems to have been twopence per candle—obviously for wax candles, as twopence in those days was a positive luxury price, despite the fact that one chandler attempted to charge fourpence. (Even as late as 1811, however, common dips were being sold at 10 to 11 shillings a dozen.)

With only the luxury candles of beeswax in competition, tallow dips held their own for common use until practically the end of the eighteenth century, and in rural areas well into the nineteenth. Local manufacture went on until comparatively recently—e.g. in Cornwall, where tallow dips were used not long ago in the tin-mines and in the Somerset coal-field, where the collieries sometimes made their own for use in the gas-free mines of the Radstock area.

Cornish tin-miners attached their dips (first tallow and later of paraffin wax) to their leather caps by means of lumps of clay, the Radstock miners wearing theirs also on their caps but held in a special clip.

Tallow dips are still produced commercially for certain purposes, as where a large flame, not easily extinguished by draughts, is required; large quantities are exported to the West Indies, but this fact takes us rather outside the realm of lighting, since the coloured folk are said to use them there for anointing their bodies in order to keep off the mosquitoes! (This seems a rather strange echo of the use of man's first light, the fire, to keep larger animals away!) On modern production lines tallow dips are made by means of frames which enable about a gross of twisted cotton wicks to be dipped simultaneously, the wicks being strung on rods holding twenty-four apiece; after each dipping they are raised and set aside to cool while another frame is dipped, then redipped, and so on until, by means of a balanced weight, the workman ascertains that the required thickness has been attained. (Snuffless dips, having cotton wicks, have now taken the place of tallow dips for most purposes and are composed of fatty acids and low-grade paraffin wax.)

The first departure from the centuries-old alternatives of tallow and beeswax came with the opening up of the sperm-whale fishery in the eighteenth century. Spermaceti, a crystalline substance from the head of the sperm whale, came into use for candle-making near the end of that century, the crude material being separated by several filtrations and pressings and purified by chemical treatment. Spermaceti candles became, and are still used as, a standard measure of artificial light; the term so often used, one 'candle-power', being based on the light given by a pure spermaceti candle, weighing one-sixth of a pound, burning at the rate of 120 grains per hour. Spermaceti candles usually contain a small percentage of beeswax to counteract the crystalline structure of the raw material and are moulded in frames.

The next step was the manufacture of stearine—fat from which the smoky and smelly glycerin has been extracted. This was the outcome of Chevreul's researches up to 1823.

Two years later (1825) the old twisted strands of cotton previously used as wicks (and still so used in dips) were challenged by the plaited wick invented by Cambacéres, which sounded the death-knell of the snuffers, since the automatic turning-over of the burnt tip of the plaited wick exposed it to the air and caused it to be self-snuffing. In 1840 the firm of Price's introduced it into 'snuffless composite candles' first used in the illuminations for Queen Victoria's marriage, and its superiority became generally accepted. (In the meantime the manufacture of candles in England was assisted by the repeal, in 1831, of a tax imposed in 1709.)

Fifteen years afterwards (1855) M. de Milly effected important improvements in the lime purification of stearine, which remained in vogue until the Twitchell process of steam-boiling the fat in a large wooden vat with a small percentage of sulphonated fatty acid compound came into being, in comparatively recent times. Stearine was at first produced from tallow, and later from other fats, including palm oil—the use of which helped to suppress the slave industry, since West African chiefs found it more profitable to utilize their subjects' labour for its collection than to sell them as slaves. Stearine burns with a clear smokeless flame, does not gutter, and is particularly suited for tropical use in that it does not bend when exposed to warmth.

De Milly's discoveries, however, were preceded and their utility somewhat discounted by the appearance of an entirely new material on the scene in 1850, after twenty years of experimental work upon it. This was paraffin wax. The middle of the nineteenth century was probably the most revolutionary period of the whole history of lighting, and the same year saw a turning-point in the twin histories of the candle and the lamp through the advent of the same basic material, paraffin. In these days of oil-prospecting in Derbyshire it is interesting to recall that it was from Derbyshire petroleum that Mr. James Young produced paraffin in the first instance on a commercial scale, a little later source being the West Lothian coal-field, and the first American petroleum discovery not taking place until nine years later—preceding those in Burma, Borneo, Poland, &c. Crude paraffin is refined for candle-making, to extract the paraffin of high melting-point suitable for the purpose; this was first done by dissolving paraffin scale in naphtha, but the sweating system (introduced by Mr. John Hodges, of Price's) has now superseded this method. Paraffin candles, which form the vast majority of those now used in the British Isles, give a greater intensity of light than stearine but are more apt to bend in a warm atmosphere.

Making in moulds, which has taken the place of dipping as the most common method of candle-making, is later in origin than the latter and is more suitable for use with textile wicks than with the fragile rush wick so often used in rural candle-making. It is said to have been introduced by the Sieur de Brex, of Paris, in the fifteenth century. Early moulds were wooden frames surmounted by a filling-trough, the wicks being provided with loops which were drawn through holes in the tips of the moulding tubes and attached to a taut-drawn wire. Later, moulds were in tin or iron, sometimes single, usually in 'batteries' like a row of organ pipes, but the author possesses an iron specimen in which the tubes or cylinders are grouped in a circle and supported on a stand.[1] Moulds were sometimes used in combination with dipping, to shape the candle after making; this method was employed quite recently by the Radstock colliery, who made their own dips.

In modern candle-making the moulds are attached to hollow piston-rods, tips downwards, and are in a tank. After being steam heated, they are filled with molten wax and the tank is filled with cold water to expedite the cooling of the wax. Wicks are supplied through the piston-rods from spools contained in a box underneath the frame. The turning of a lever causes the piston-rods to force

[1] Plate I, No. 9.

the candles from the moulds; they are then supported in a clamp until the completion of the next moulding, when the wicks are severed and they are removed.

Beeswax candles continue to be manufactured for use in Catholic churches, owing to a Papal Bull which, in order to exclude animal fats from the altars in a day when tallow was the only alternative to beeswax, ordained that candles for religious use must contain a high proportion of the latter material—candles for the High Altars having 65 to 75 per cent. and those for lesser altars at least 25 per cent. These, owing to the wax adhering to the sides of moulds, were and are made by the pouring method. The wicks are attached to a wooden hoop, suspended over a cauldron of wax, and the workman pours the material from a ladle on to the wicks while rotating the hoop. After a certain thickness is reached the candles are inverted and the pouring continued; finally they are laid, in a plastic state, on a marble slab, rolled, and cut. A print in the possession of Messrs. Field shows precisely this method in use in 1749 and it was probably then already old.

The catalogue of the Smithsonian Institution (U.S.A.) gives an interesting summary of candle-making in China and Japan.

'The first requirements are slender rods finished from bamboo and tapered from base to point. The next is a spill or tube of paper wound spirally, with rush pith forming the wick. This is slipped on the bamboo rod. The Chinese thus solved the problem of the capillary wick. The combination is then dipped in the melted wax or fat and cooled alternatively until a candle of the calibre required is secured. . . . The pink candles of graded size are made by the same method as are the large candles. In the latter, a vegetable stem is used instead of a bamboo rod, upon which the rush pith wick is wound.'

The Japanese also moulded candles in paper tubes. Chinese and Japanese candles were frequently decorated or ornamented with characters. The materials used include wax from the Cocus (wax insect) and from the seeds of certain trees.

In the United States the fruit of the bayberry was used for candle-making by the early settlers, as well as the fat of deer and bears and wild bees' wax. The bayberry provided the choicest candles, but the process involved a series of boiling and skimming before the fat became the characteristic semi-transparent green; the candles made of it burn with a fragrance certainly foreign to the tallow candles. They are still made and sold. Unlike the old country the States made a greater use of oil lamps (largely apparently of Dutch origin)—especially in the eighteenth century—and candles remained a luxury for many years. In 1634, according to Mr. Hayward in *Colonial Lighting*, no candles were to be had at a less price than fourpence each, a figure which of course meant a great deal more in value than the same price to-day. A great many candles, however, were either home-made or made, as in medieval England, by itinerant candle-makers.

In the East as well as the New World vegetable materials have been utilized

PLATE V

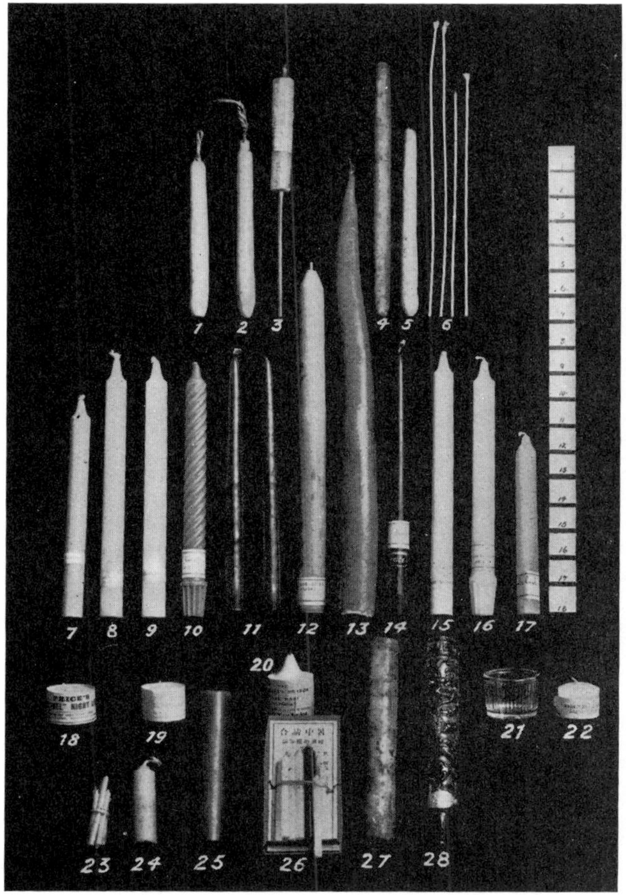

(*a*)

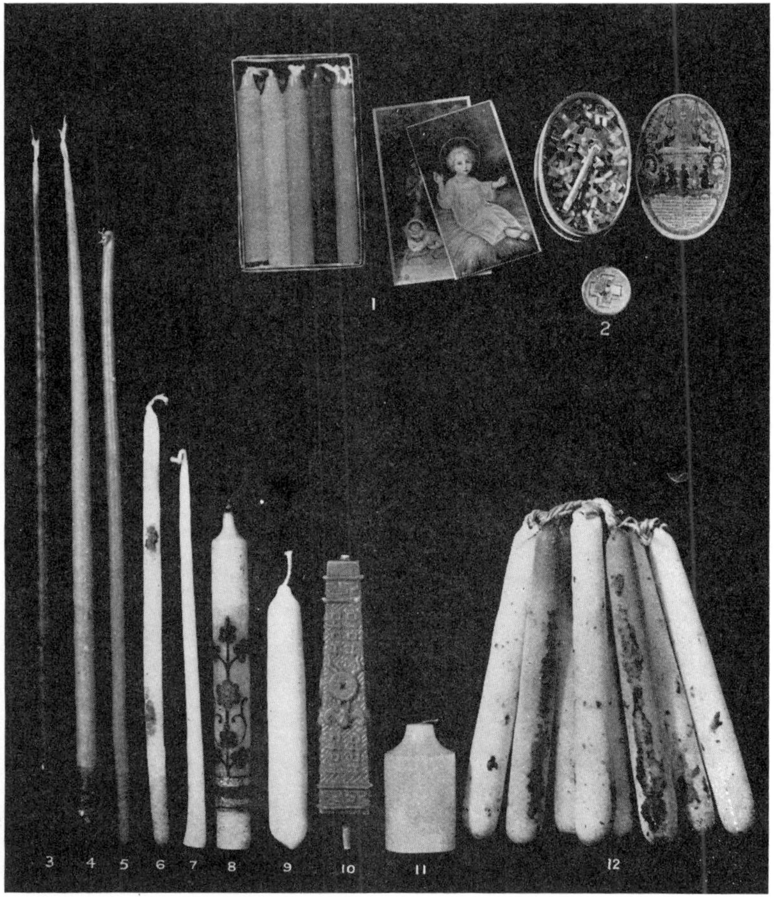

(*b*)

DIPS, CANDLES, AND NIGHTLIGHTS

for candle-making apart from the cases given, as for instance the oily seeds of the candleberry and the tallow tree in the East, the waxberry and birch bark in South America, and myrtle wax in parts of the West Indies.

Something must be said here as to night-lights; these, a frequent appointment of the nursery and the sick-room not many years ago but now less frequently used, are practically short squat candles made to burn for six to ten hours and are said to have originated with the 'mortars' or mortuary candles burnt in death chambers, which, however, were more akin to tapers. The three main forms, usually made of paraffin wax, are (*a*) in paper cases, to burn in a saucer with water, (*b*) in paper cases, to burn in a saucer without water, and (*c*) without cases, to burn in small glasses. The earliest form in cases (introduced by Mr. Childs in 1849) was formed by pouring semi-fluid fat or wax by hand into the case already provided with a wick and 'sustainer', a method now almost superseded by moulded night-lights having the wick fixed separately through a perforation partly formed in the moulding process and completed by means of a heated needle. A fourth type, the 'Pyramid' night-light, introduced by Samuel Clarke in 1857 and still manufactured, is of hard neutral tallow with a fine rush wick and a conical candle-like tip; this, too, is moulded, a paper case being wrapped around and the bottom closed with plaster of Paris. Burning as it does with a larger flame, it is a type especially favoured for food warmers and under shades.

Tapers, in the *modern* sense of the word, such as are made in large quantities and sold in Catholic countries for religious processional purposes, are slender candles intended for carrying in the hand. Coloured tapers are used in festivals such as at Christmas—hence no doubt the coloured candle of the Christmas tree, just as the candles on the birthday cake may be an echo of the medieval practice of placing a taper in the hand of a child at baptism.

[*See Plates I, IV, and V.*]

CHAPTER V

THE CANDLESTICK

It is curious how, in many respects, Roman civilization and implements are much nearer to those of the present day than were those of the Dark and Middle Ages. The candlestick is a case in point. While a 'spike' or 'pricket' is a feature of Etruscan candelabra, the Roman candlestick is usually a socketed arrangement, very closely approaching the candlesticks (particularly the ceramic varieties) of modern days. Yet the medieval candlesticks are entirely of the pricket type for the earlier part of the Middle Ages and mostly so for the latter part, while the pricket form survives to the present day in the tradition-governed furniture of continental churches.

The Hebrew 'seven-branched' candlestick was not a candlestick at all but a group of lamps, while the candlesticks of the Tut-ankh-amen tomb are more properly to be classed as torch holders. The Etruscan candelabrum, with its branching spikes, approaches more nearly to the idea of a candle implement, but its 'candles' were more likely to have been 'splinter' lights; the Roman candelabrum, based though it is on the Etruscan form, has its upper end terminating in a way which brands it as a lamp-stand, so that its description as a 'candelabrum', in the modern sense of the word, is misleading.

Neuburger, in *Technical Arts of the Ancients*, refers to Roman candlesticks with spikes or sockets, and, according to the British Museum 'Early Christian' guide,[1] candles and tapers in the catacombs were set either in sockets or on prickets. A combined pricket and socket candelabrum found in the river Witham, at Kirkstead Abbey, has been described as Roman, but there seems to be some room for doubt about the ascription—the form is distinctly suggestive of the sixteenth or seventeenth century, a period when the combination was not uncommon. There are, however, pricket candlesticks of Roman date in the British and Guildhall Museums.

Generally speaking, Roman candlesticks found have been of the socket variety, whether in metal, clay, wood, or, occasionally, stone. An interesting series was found in the excavations at Wroxeter (Uriconium), the general form being more or less that of a goblet, with splayed foot and bowl top, the latter having a central socket for the reception of the candle; a somewhat similar form is made in pottery in Italy at the present day. The Wroxeter examples are in grey or red clay, with a white or cream slip. Silchester yielded candlesticks of terracotta ware with 'stems' and 'feet' and a central band; some were provided with midway collars after the style of Tudor examples, including concave forms for drip-catching. The flat 'bedroom' candlestick type was represented there and in London in the form of a shallow bowl provided with a central

[1] *Guide to Early Christian and Byzantine Antiquities.*

socket. Silchester also provided metal candlesticks of various forms, including one consisting of four sockets joined together at their bases, so that when it was set on a flat surface one socket only was in a perpendicular position to receive the candle, and others in the form of brackets composed of flat bars, each with a socket at one end and a spike to be driven into the wall at the other. A Roman candlestick in the Guildhall Museum shows the spike principle of fixing the implement in a vertical form.

The Silchester four-socket candlestick has a relation in a two-ended variety recorded by Neuburger as having been found at Salzburg; this is said to have been provided with sockets of different diameters, so that candles of different thicknesses (obviously crude dips) could be used. A post-medieval development is foreshadowed in this example by the provision of an aperture to assist the removal of the candle-end, a device which also appears in the socket of a flat pottery candlestick from Rheims in the British Museum—catalogued as a lamp-stand. In the same way, the central collars of the sixteenth- and seventeenth-century candlesticks are anticipated both by the Silchester examples already cited and by collared candlesticks in the London Museum.

There is a break between the Roman candle-holder and the development of the candlestick in later days. The very name 'candlestick' is eloquent of its beginning—as a spike, no doubt a pointed stake (as in the case of the split-wood rushlight holders) on which the candle was stuck, all idea of it being set in a socketed receptacle having been lost or abandoned. The pricket type, as mentioned above, had been in use in Roman times and was used side by side with socket candlesticks in the Early Christian church; its very form suggests a primitive beginning and its prevalence in the Dark and Middle Ages is eloquent of a reversion to crude barbarism.

There are references to candlesticks in Saxon times, but the earliest extant or depicted, to the best of the author's belief, date from the eleventh to the thirteenth centuries, these being of the spike or pricket type; there are good examples portrayed in one of the stained glass windows of Bruges Cathedral. These early candlesticks were generally of iron, bronze, or latten (medieval brass); the well-known Gloucester candlestick (twelfth century) in the South Kensington Museum is of silver-gilt. They are sometimes ornamental, as in the case of the Gloucester example, but frequently the rough production of the local smith. The commoner varieties often terminate at the bottom end in three or more feet, strongly suggesting a tradition carried on later by rushlight holders, whose workmanship is of the same type of craftsman. The University Museum at Oslo contains interesting examples of medieval pricket candlesticks in iron and bronze which suggest a Teutonic origin for those found in this country.

The pricket candlestick in various forms remained paramount during the greater part of the Middle Ages. As a single candlestick it ranged from the simple iron spike surmounting a plain shaft on a flat base or crude feet to elaborate examples decorated with Limoges and other enamels; there were also many branched candelabra for standing or hanging, coronae or chandeliers for varying numbers of candles, and grotesque forms such as dragons or the like holding the prickets for the candles on their backs. (A bronze example in Southampton Museum is in the form of a stag with a pricket on its back.) Most of the pricket candlesticks which have survived are of ecclesiastical origin. Several interesting and ornate examples were illustrated in the *Connoisseur*, vol. x, no. 38; these include specimens from St. Paul's Cathedral, Antwerp Museum (a wrought-iron wall-bracket variety), Magdeburg Cathedral (standing), Lubeck Cathedral (hanging, wood), Wurzburg Cathedral (standing, coronet type for several candles), and St. Lorenzo, Rome (pricket set on a marble column). Noyon (France) boasts of two fine examples of decorative ironwork in pricket candlesticks—one a paschal candlestick in the medieval hospital and the other in the cathedral. The materials used were iron, copper, or bronze generally, but wooden specimens are not unknown.

There are a fair number of examples of old pricket candlesticks in German museums and church treasuries. Ecclesiastical pricket candlesticks sometimes took the traditional form of the seven-branched candlestick, as in the case of examples at Essen Minster (eleventh century)—in bronze—at Brunswick Cathedral (14½ feet high), and brass examples elsewhere. Late Gothic examples with three to seven branches are in several of the German churches; the Marienkirke, Danzig, has bracket types in pairs side by side and others with three prickets arranged in a line from front to back. An interesting example at Urnaes church (Norway) has a line of nine prickets arranged on a stand in the form of a Viking ship.

Pricket candlesticks lasted in this country into the sixteenth century, with isolated examples even later. A Charles II wall panel of gilt plaster on lime wood, carrying a bracket, and lately in the possession of Mr. Sharp, Winchester, had an iron pricket for the candle. In many cases late examples are combined with sockets in one candlestick or candelabrum. (A possibly accidental survival of this sort of thing is the provision in comparatively recent rural candlesticks of a spike or 'save-all' on which to stick the candle when it has burnt too low for the socket.) The re-advent of the socket candlestick in the fourteenth or fifteenth century, however, caused a decline in the pricket variety and socket candlesticks are more common from the sixteenth century onwards.

On the continent of Europe ecclesiastical tradition has preserved the pricket device in churches until the present day, both on single candlesticks (usually in brass, but sometimes in pewter, wood, and other materials) and on stands for votive candles, which latter are usually in iron and take either the form of horizontal bar-holding prickets in a line or in superimposed rows (pergulae or espaliers, herciae or harrows, and rastella or rakes) or that of a circular 'corona'. An early English example of the former group, as a wall-bracket arrangement, is in Rowlestone Church (Herefordshire)—thirteenth century—and the latter

(*a*) PRICKET CANDLESTICKS

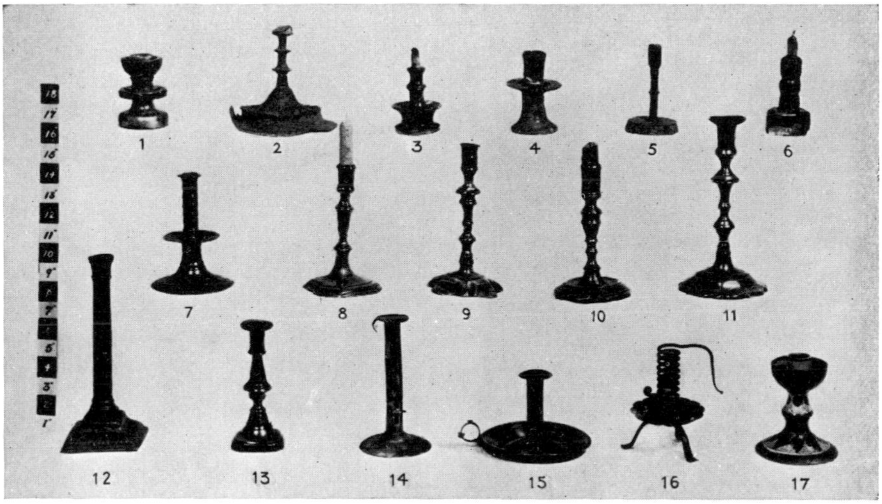

(*b*) SOCKET CANDLESTICKS

is a close relation of the medieval 'trendal' or 'roundel'—a circular band with prickets set on it, suspended before the Rood (a number of examples of which are to be seen in Norwegian churches). Many of the continental votive candelabra are old and some of them are fine examples of ornamental ironwork; the greater number are plain. Notable specimens of the ornamental type are to be seen in some of the churches of Belgium, northern France, Germany, and Scandinavia.

The 'roundel', in some cases, took an elaborate form which entitled it to be classed as a chandelier. An eleventh-century example at Hildesheim, 20 feet in diameter, boasts 72 candle-holders and a number of 'towers' and statuettes. Aix-la-Chapelle Minster possesses one 13 feet in diameter with 48 holders, with engravings of the beatitudes and scenes from the life of Christ in pierced work. A twelfth-century example at Comburg Monastery, Swabia, combines statuettes with foliate and animal ornamentation.

The pricket candlestick is also found in eastern Asia and is the prevalent type in China and Japan, usually in bronze, brass, porcelain, or pewter. Ceremonial sets were used in ancestor worship and the elaborate altar specimens frequently incorporate human or mythological figures. Those for more ordinary use are simpler. Most of the Asiatic varieties show an advance in utilitarian design over the early European specimens in that they are provided with a bowl or saucer at the top, in which the pricket is set, thus preventing grease running down the stem. Some of the European examples have this refinement, but it is absent, not only in the earliest specimens, but often in the simpler candelabra for votive candles also, with the result that continental chapels are frequently disfigured with candle grease.

Putting aside the Roman candlesticks, the earliest European socket candlesticks are probably of the fourteenth century. The idea seems to have come from the East with the wax trade, via Venice, as socket candlesticks were in Persia at least as early as the thirteenth century and many of the early European forms of the type betray Oriental influence.

So far as England is concerned, two portable candle-holders shown in the Luttrell Psalter (fourteenth century) look almost as though they are socketed, but as they are shown carrying candles it is hard to say. The well-known Beaufort candlesticks (wood) at Winchester are socketed and of the fifteenth century. Continental examples have been dated to the fourteenth century, including some which carry on the form in which an animal holds the candle receptacle on its back. *A Description . . . of all the Ancient Monuments . . . within the Monastical Church of Durham before the Suppression*, written in 1593, gives details of a Paschal Candlestick and other notable candlesticks and candelabra which appear to have been socketed (certain iron sticks are distinctly given as such), and these must have been installed at least by the end of the fifteenth century in all probability, since the dissolution of the monastery took place in 1539.

THE CANDLESTICK

Although the wooden Beaufort candlesticks show the midway collar common at a later period, the earliest socket candlesticks in metal (usually iron, latten, or brass and bronze) are mostly very plain uprights, terminating at the base with three feet or a disk resting on three feet; their sockets are open sided, to facilitate the removal of the candle ends, and this arrangement, gradually developing into the provision of a slot in an otherwise closed-in socket, lasts to the end of the sixteenth century. Many of the socket candlesticks ascribed to the close of the Middle Ages are provided with projecting arms which, it is suggested, were to enable them to be suspended if desired. The pricket idea died hard; not only did it last long after the introduction of the socket candlestick but, in many cases, it was combined with the socket device in one candlestick.

In the fifteenth century a type of base appears which is strongly suggestive of an eastern origin; this is analogous to a type known in Persia in the thirteenth century—a high circular base surmounted by a drip pan—and spread over the European continent (more so, apparently, than in the British Isles) even as far north as Norway. It lasted, in a modified form with curved instead of straight sides, through the sixteenth century and into the seventeenth.

It is interesting, from this point, to note the upward progression of the drip-catching arrangement throughout the development of the socket candlestick until (after a lapse in which it seems to disappear altogether, or remain in a vestigial form only, during part of the eighteenth century)[1] it finally finishes up as a rim at the top of the stick as at present.

In the sixteenth century the drip-pan first becomes a collar midway up the stem. In this century there is a marked increase in the use of the socket candlestick. Apart from metal types a number of pottery examples (partly or wholly glazed, in green or yellow), with the collar, are extant, a considerable number having been found in London (many of which are said to have come from the potteries at Farnham, Surrey). By this time there had developed a somewhat greater tendency to depart from the straight stems of the earliest metal candlesticks in favour of 'baluster' forms and brass was rapidly replacing iron as the common material. Precious metals were seldom used for candlestick manufacture until the reign of Charles II, but, from then on, silver was the usual material for better-class candlesticks (brass for the commoner).

By the seventeenth century the square slot for removing the candle end had been replaced by a small hole on one side of the socket only. The midway collar remained until the latter part of the century, when it was replaced by ornamental 'bulges' or curves and, where retained as an integral part of the stick, took up its present position as a much diminished feature at the top of the socket (as a flange).

During that century an important alteration took place in the method of making brass candlesticks which, hitherto having had a solid cast or turned stem attached at the base by screw or tenon (usually the latter), henceforward

[1] A separate detachable rim was sometimes provided.

were made with hollow stems, the stem and socket being cast separately, brazed together and subsequently attached to the base. The provision of a rod running through the hollow stem to expel the candle end spelt the end of the aperture in the side of the socket.

The eighteenth century was a period in which almost every decade had its special fashion, ranging from the elegant examples of the first half, with octagonal or floriate bases and baluster stems, to the fluted column and square-based types of the latter half of the century, when the craze for things classical or pseudo-classical extended to candlesticks as to so much else. Sheffield plate then figured as an additional metal in candlestick manufacture, as well as pewter. The glass candlesticks of the eighteenth century, so much in demand by present-day collectors, followed the general lines of the silver, Sheffield plate, and brass types and were luxuries rather than in general use.

Wooden candlesticks are, by reason of their inflammable nature, seldom used except in poor or remote areas, but are favourite objects of peasant-craft in northern Europe. The Northern Museum at Stockholm shows a large number of examples, the most typical of Scandinavian villages being a type in which four curved supports of fantastic design hold a candle socket in their centre. Finland and Hungary are other fairly prolific sources, and there are examples in Scottish museums; obviously wooden candlesticks are commoner in parts where wood is plentiful and metal less so.

Iron candlesticks of the socket type, as used in country districts in particular, are generally in two forms—the dip holder, either in combination with the rush-light holder or independently, and the hollow cylindrical candlestick. In the former some of the early examples have merely a ring or collar to grip the dip; some Norwegian examples pinch it in between two 'spring' pieces of iron. These would appear to be prior to the socket in cultural sequence. More usually the crude dip holder is in the form of a socket like an inverted cone, made by hammering over the iron.[1] The cylindrical candlestick in its commonest form has a shallow inverted cup base which, it is said, was used in this country for scraping hogs at butchering! This candlestick (which seems to be mainly or entirely English) was usually provided with a hook at the rim which enabled it to be hung on a shelf or projection; it always has the sliding 'spur', working in a slot, which was a method of raising the candle introduced probably in the eighteenth century and common in the ordinary candlesticks, iron or brass, of the early nineteenth century. An earlier method of securing the same result had been the 'lift' candlesticks of the Jacobean era (copied in iron 'coil' candlesticks of later times) in which the 'lift' progressed in a spiral slot running up the stem of the iron band candlestick. Another idea of this sort is present in a type common on the Continent, in which the 'stick' takes a cage-like form of vertical iron bars having the socket inside, elevation and lowering being by sliding up and down the outside bars. This type usually has a drip pan immediately above the base.

[1] Romano-British iron sockets are strikingly similar to those of the 18th–19th centuries.

THE CANDLESTICK

Porcelain candlesticks belong to the last two centuries or so, but pottery candlesticks, both of the flat and upright types, are widespread and sometimes take very primitive forms. A by no means uncommon type in Scandinavia recalls the Roman candlestick with its four sockets and reversible positions; this is generally in coarse earthenware and a peasant production. Some North African candlesticks retain the midway collar of the seventeenth century in Europe. Usually pottery candlesticks reflect the pottery industry and the style of decoration of the country to which they belong.

Candelabra, Chandeliers, &c. No doubt the chandelier or hanging candelabrum includes in its ancestry the 'polycandelon' or hanging disk for glass lamps of the Byzantine era. Its family tree passes through the trental or hanging circlet for prickets of the Middle Ages. So long as candles remained the great illuminating 'standby', some such device was necessary for lighting extensive interiors, and it is natural that most of the extant hanging candelabra are ecclesiastical in origin. A very fine brass example in the Temple Church, Bristol, dates from the late fourteenth century. A few other churches possess sixteenth-century examples, but the majority are seventeenth or eighteenth century in date; St. Mary Redcliffe, Bristol, and Winchester Cathedral are amongst the numerous English churches possessing specimens of these later periods. The type varies little here or elsewhere, the composition being that of a central globe carrying a varying number of 'S'-shaped socketed branches, and the material being brass.

Glass chandeliers belong generally to the eighteenth century and, although one example is recorded as having been in Thomastown Parish Church, Ireland, their homes were more usually palaces and mansions.

The halls and churches of the period, with their scores or hundreds of candles, must have been uncomfortably warm in times of assembly and one does not wonder at the tendency of Georgian ladies to faint on every possible (and convenient) occasion!

The candle is a very adaptable article. Hence candlesticks, portable, stationary, and in chandeliers, were reinforced by brackets, sconces, and all sorts of devices capable of attachment to wall, ceiling, or anywhere else, which might hold a socket (or, earlier, a pricket). To describe or collect every possible variety of candle-holder would be an endless task and one would have to include wine bottles, ordinary saucers, and (in Cornwall, for instance) sea shells.

[*See Plate VI and frontispiece.*]

SECTION II

THE LAMP ITSELF

CHAPTER VI

THE LAMP OF STONE

THE lamp proper began when, somewhere in the Old Stone Age, man (or men—perhaps even women) discovered that a wick, almost certainly of vegetable origin, soaked in and fed by fat, provided a lasting light. No doubt this discovery was accidental, whether it was single or, as is more likely to have been the case, multiple. As the author suggests in his story, 'The Lamp',[1] probably some primitive man, idly watching the fire at the cave mouth, gradually noticed that some vegetable or animal substance (a piece of moss, as used by the Eskimos, a dry reed, a lump of twisted hair) had dropped into a pool of fat from the carcass which had dripped into a hollow stone standing by the edge of the fire and was burning steadily and independently. Curiosity would make him draw the stone away—and possibly burn his fingers in his haste to examine the phenomenon. Finally, he would make the momentous discovery that this little flame could be carried about in the interior of the cave and would dispel the spirit-haunted shadows far away from the hearth. Unlike the torch, it could be set down readily and did not burn away so rapidly.

This was how the lamp was born, at any rate in western Europe and North America and possibly in the interior of Asia, and until there is definite evidence of the use of sea-shells as lamps in the Stone Age the primitive stone lamps of Palaeolithic man hold the field as the first real lamps in history.

The birth of the lamp (since the torch was at best an expedient of dire necessity) no doubt heralded the beginnings of night life. The sophisticated *habitué* of the twentieth-century night club little realizes what he owes to the inquisitiveness and even the reasoning powers of the earlier barbarians; no doubt the elders of the cave held up their hands in horror (if they did not act in more drastic fashion) when the young bloods of the period defied convention and the rule of the family by staying up playing knucklebones until eight or nine o'clock of a winter night, with the aid of the new invention!

Naturally enough, the earliest stone lamps were crude and it is difficult to identify them with certainty. By degrees, Dame Utility suggested improvements, such as a handy shape for the lamp, a groove for the wick, a slope towards the wick in the bowl itself, and a resting-place for the hand away from the flame. The last-mentioned idea is decidedly suggested by a stone lamp from Cornwall in the author's possession, found in conjunction with polished stone axes of the close of the Neolithic period; this not only has a platform for the thumb on one side of the lamp (wider at the end away from the wick, as it should be) but also a hollow on the under side which receives the fingers in

[1] *Daily Express*, 10 Nov. 1934.

easy fashion.[1] It is rather curious to note that this lamp has one straight side and one bowed, thus conforming roughly to the much later and more finished stone lamps of the Eskimos. It is 5½ in. long × 2¾ in. wide. From the same part of Cornwall (West Penwith) came a diamond-shaped stone lamp with a projection at one corner as a handle, in the possession of Mr. Noall, St. Ives (to whom the author is indebted for his own), as well as other practically certain lamps.

These lamps are comparatively late in the history of the stone lamp so far as actual dating is concerned. Mr. Norman Ault (in *Life in Ancient Britain*) suggests the Magdalenian period (perhaps some 15,000 years ago) as producing the first lamps, but recent discoveries suggest an even earlier advent.[2] Perhaps the most significant of these are the finds made by Dr. Karl Absolon at Pekarna, Moravia, which he dates as something like 30,000 years old. These include oval and circular bowls of sandstone (which, no doubt because of its easy working, seems to have been a favourite material for lamps), 8 to 20 cm. in diameter and artificially hollowed. Some of these may have been used as mortars for pounding dyes (since a piece of red chalk was found ground into one), but some of them are almost certainly lamps—possibly they were used for either purpose, just as some prehistoric urns might have been used as kitchen utensils as well as for cremated burials (a not unnatural, though rather gruesome, association of ideas!). An ear-shaped sandstone vessel from the Pekarna excavations is particularly interesting, as its form is very close indeed to that of the author's prehistoric lamp from Cornwall, though without the 'wick groove', a feature which appears to be lacking in the Pekarna lamps generally and is no doubt, and naturally, an improvement on the original idea—an improvement, incidentally, which does not seem to have occurred to the Eskimos.

Neolithic (*c.* 4000 B.C.) finds at Khirokitia, Cyprus, however, include a number of stone bowls, some of which may well be (and probably are) lamps, and these include specimens with distinct spouts, even more advanced and pronounced in form than the wick groove of the Cornish specimen. Handles (horizontal and flat or 'vertical' and curved) appear on a number of these bowls and an attempt at decoration on some of them.

The well-known chalk lamp from Cissbury, near Worthing, is usually regarded as belonging to the transitional period between the Old and New Stone Ages (say about 10,000 B.C.). The flint knappers of Brandon, Suffolk, true to the setting of their industry, used a chalk lamp such as the example now in the Smithsonian collection (U.S.A.), in modern times. Another chalk lamp, in Devizes Museum, from Knap Hill excavations, 3 in. × 2 in. in size, is classed as Romano-British.

[1] Plate VII (*a*), No. 3.

[2] A round sandstone lamp of the Magdalenian period, found in the Grotte de la Mouthe (Dordogne) and preserved in the very interesting museum at S. Germain-en-Laye, near Paris, has a very definite spout and is engraved on the base with the head of an ibex.

(*a*) STONE LAMPS FROM THE AUTHOR'S COLLECTION

(*By the courtesy of the Science Museum*)

(*b*) STONE AND SHELL LAMPS IN THE SCIENCE MUSEUM, SOUTH KENSINGTON

The hollowed-out stone persisted in its simplest forms as a lamp throughout the prehistoric eras. In the anthropological collection of the Marischal College, Aberdeen University, are a number of hollowed stones classified as 'cups', in steatite, granite, and sandstone, ranging from $2\frac{3}{8}$ in. to $4\frac{3}{8}$ in. in diameter and 1 in. to $2\frac{1}{2}$ in. in height, some with unperforated handles; these are strongly suggestive of primitive lamps. We are on more certain ground with some stone lamps of the Early Iron Age in the Museum of National Antiquities of Scotland. These show two circular depressions running into one another as reservoirs—the larger, it is suggested, was for the oil, the smaller for the wick; some indication of the same duality of the hollows is shown in the prehistoric lamp from Cornwall already cited. The Scottish examples are from the briochs of Kettleburn and Okstrav (Orkneys) and are in sandstone; a more finished specimen is from an underground dwelling at Tealing, Forfar. Apart from such stone objects definitely classified as lamps, the Museum possesses, like the Aberdeen collection, stone 'cups' which may also have been lamps, and these, too, came from Aberdeenshire. The occurrence of such a number of possible lamps in that county suggests that, since the lamp was certainly more economical than the torch for constant use, the Aberdeen tradition had already taken hold!

Evans, in *Ancient Stone Implements of Great Britain*, records several probable lamps, e.g.

> Six or seven stone vessels of various sizes and forms discovered in a kist-vaen in the island of Unst, four being of a rough quadrangular form with flat bottoms and from $3\frac{1}{2}$ in. to 7 in. in height, the other three oval. These were of schistose rock and some bore traces of fire.
>
> Several similar vessels from Skara, Orkney.
>
> Two stone 'cups' or hollowed-out stones from Ty Mawr (one illustrated certainly appears to be a lamp. F. W. R.).
>
> Some small cup-shaped vessels of chalk, probably used as lamps, found by the Rev. W. Greenwell, F.S.A., at Grime's Graves.
>
> A cylindrical stone vessel, 5 in. high $\times$ $6\frac{1}{2}$ in. diameter, with a cup-shaped cavity above and a small hole below 'as if for fixing it on a stand', from Parton, Kirkcudbrightshire. (Note the hole or hollow below—cf. the author's Cornish specimen.)
>
> Another stone vessel, circular in form and ribbed externally like a melon, found with a polished stone hatchet in a cairn in Caithness.

All these appear to have been of Neolithic date, but another example, of later date, was found in a barrow at Wetton with an iron-handled bronze bucket. This was 4 in. high, of sandstone, with four grooves around it as ornament.

The chalk lamp from the Knap Hill excavations in the Devizes Museum given as 'Roman' may be correct as to period, as there is a stone lamp from Winchester in the Tudor House Museum, Southampton, which was found at the same level as Roman objects, and a square stone lamp among Roman objects in the Guildhall Museum, London. A square stone lamp from Wookey Hole,[1]

[1] Wells Museum.

in the Mendips, is also probably not much older than the period of the Roman occupation. All these cases, however, probably represent a native tradition persisting in spite of superimposed culture, notwithstanding the presence of stone lamps on the Acropolis and at Mycenae (*vide* British Museum catalogue). The stone lamp in the Near East and Mediterranean areas indeed seems not to be so much in the line of direct succession as the adoption of an alternative material for lamps which had already been evolved from a different origin. In this way, calcite was used for lamps at Ur which were undoubtedly derived from the sea-shell in the first instance; and later, stone open lamps of pronouncedly advanced forms appear in Egypt (Arab in period) and Persia. Occasionally stone lamps have appeared in true classic forms: the British Museum list five from Greek and Egyptian sites, obviously adaptations from pottery forms.

A number of early Cretan lamps in stone, however, show a technique all their own, but bear little resemblance to the primitive hollowed-out stones which form the beginnings of the lamp. These Cretan lamps are shallow decorated bowls of considerable size, set on pedestals, a red gypsum example in the British Museum being 2 ft. or more high. Another, lower, is in black steatite and dated as far back as 1500 B.C.—contemporary therefore with the early shell-form pottery lamps of the island. No doubt the latter were the ordinary domestic utensils and the stone lamps for temple and palace illumination.

Primitive stone lamps have appeared in other parts of the world spasmodically, especially in comparatively modern times, among backward races or in places where tradition outlives its period. The Smithsonian Institution (Washington, U.S.A.) show one from Baltistan strongly reminiscent of three from Egypt known to the writer—one a pointed oval form, the others in a boat-shaped form with the nozzle at one end and an upright stub handle at the other; they appear to be early Arab. The Smithsonian collection also exhibits a number of stone lamps of circular form from Hawaii which have strongly marked affinities with some of the prehistoric stone lamps of western Europe. 'J.R.A.', in the *Journal of the Scottish Society of Antiquaries*, records that the churches of north-east Asia had lamps, presumably of stone, trimmed with wooden pins which, when soaked with oil, were used as torches.

For the most complete record of the stone lamp, we must turn to one of the farthest corners of the earth, where the Stone Age has lasted until our own day. It begins, again, with crudely hollowed stones; in the Smithsonian collection are rough beach stones, one at least quite unworked, used as lamps among the Aleuts of Alaska. These are possibly the forerunners of the Eskimo stone lamps of the present time. In southern Alaska the primitive lamps are of hard stone, sometimes of sad-iron shape and sometimes circular. They seem to run about 5 in. × 4 in. in size. Farther north, the Eskimo lamps are in soapstone and, being used for cooking and heating as well as lighting, are often very large, the wicks of powdered moss being laid along one edge, in contact with the fuel (blubber). That these Eskimo lamps have a long history behind them is shown by the discovery, by Dr. Mathiasson in Greenland (vide *Antiquity*, June 1935), of lamps of soapstone, with wick ledges, dated as medieval, some of them showing elements of decoration. More striking still were the discoveries at Kachemak Bay, both because the lamps were found there in a prehistoric village and because they evidenced a far more advanced artistic conception than any of the later stone lamps in that part of the world. Some of them showed, not only well-finished bowls, but human figures carved in the centres of the bowls—all quite unlike any lamps used by the Eskimos in historic times. They were uniform in finish, of fine-grained igneous rock, and the figures were of Eskimo type (though five of the lamps were found outside present Eskimo territory). Another lamp from the same area has two whales carved, in place of the human figure. The lamps were oval, from 8 to 16 in. long × 7 to 14 in. wide × 3 to 6 in. high, and were depicted in the *Illustrated London News*, 21 October 1933.

Archaeologists generally maintain that there is no indigenous lamp in America south of the Eskimos, but the author possesses a stone article which is unmistakably a lamp and equally unmistakably American Indian in ornamentation (it was said to have come from Peru). (Plate VII, *a*, No. 4.)

The stone lamp crops up again in the Old World during the Middle Ages in the form of the cresset, used particularly for the lighting of cloister walks and sometimes for the churches themselves. Essentially the cresset stone is the primitive stone lamp in its simplest form except that in most of the medieval examples the wick seems to have been supported (either on a float as in the sanctuary lamps of to-day or by an upright for which a small hole is provided in the bottom of the bowl) instead of hanging over the edge. This curious link between the comparative civilization of the Middle Ages and the homes of the cavemen lasted at least until the dissolution of the monasteries, and seems to have been the usual method of lighting cloisters and dorters if not other parts of the monastic living quarters.

In Sweden the stone lamp has lasted even longer. There are quite recent specimens in the Northern Museum of Antiquities, on Skansen island, Stockholm, from peasant homes in Gothland and elsewhere.

The influence of the stone lamp (or more particularly the multiple stone cresset of the Middle Ages) is manifestly present, too, in the seventeenth-century 'tailors' candlesticks' in Edinburgh Museum, each provided with four hollows for candles and a centre hollow for snuffed wicks.

[*See Plate VII.*]

ADDENDUM

Since the above chapter was written, the author has received the following note from Mr. Gargoswara Karyap of Madras:

'There are two types of granite lamps used in South India—those fixed into walls, which are comparatively small, and a massive structure with a cylindrical base and circular top' (cf. Minoan stone lamps) '2 to 3 ft. high, permanently fixed in the earth in the common ground of villagers, around which the village folk assemble on auspicious and religious occasions and in the light of which sacred folk dramas are enacted.'

CHAPTER VII

THE SHELL AND THE LAMP

WHILE in western Europe and some other parts the earliest form of lamp was a hollow lump of stone, that essential adjunct to every form of night life had an independent start in the East and the Mediterranean area. That other origin is the sea-shell, which, though its first use as a lighting appliance cannot be traced back at present more than five to six thousand years, apparently owes nothing to the stone lamp in its inception; it gets crossed with the latter in some areas at a later stage.

So far the earliest trace of the use of the natural shell itself as a lamp comes from Mesopotamia, in which connexion it must, of course, be borne in mind that in early Babylonian times the head of the Persian Gulf extended farther north than at present, making Ur and its sister cities nearer the sea. The use of the sea-shell for such a purpose, however, was obviously fairly widespread in the Near East and the Mediterranean littoral, if not elsewhere, in ancient times. The difficulty of being dogmatic as to its early use and distribution is patent; not only are shells extremely fragile but, unless direct evidence is available, there is little or nothing to distinguish a shell lamp from any other similar shell. Therefore, one has to draw reasonable conclusions, in most cases, from the circumstantial evidence provided by the forms of early manufactured lamps, backed, not only by the Mesopotamian discoveries but also by the survival of such primitive shell lamps until quite recent times in several maritime areas.

First, as to the Mesopotamian evidence: Sir Leonard Woolley found lamps at Ur which, in some cases, he dated back to the fourth millennium before Christ, and most, if not all, of them have some direct relationship to the shell. The most primitive and therefore presumably the earliest of these lamps are actual conch shells with the natural orifices enlarged and with the whorls and columns cut away.[1] In some instances an addition of some sort was made by way of decoration, an advanced example having a stone head added to give the whole thing the appearance of a duck. The essential part of the lamp was still the shell itself.

The line the development subsequently took is indicated by the following extract from Sir Leonard Woolley's report on the 1933–4 excavations:

'An alabaster lamp, U. 19744, is covered all over with dot decoration but at one end a gazelle's head has been cut in low relief as if to break the mechanical monotony of the main design. Another alabaster lamp, U. 19745, is much more curious. The form is taken from the tridacna shell, *of which were found in the grave several examples cut open for use as lamps,*

[1] See Plate VIII (*a*). Similar finds at Chanhu-Daru, India, have been described as 'children's feeding cups', but that they were lamps seems far more likely.

and the copy, with its five projecting horns, grooved to receive the wicks, is sufficiently true to nature. But, moved by some whim of fashion, the maker has had underneath a bat's head carved in the round and, seen from below, the lamp has all the appearance of a flying bat, the horned shell becoming its ribbed extended wing.'

The italics are not Sir Leonard's but serve to emphasize the origin of the Babylonian lamp which, but for the connecting links, might be obscured by the subsequent elaborations. The tridacna is, of course, the giant clam, the form of which would naturally suggest a multiple-nozzle type of lamp.

These particular lamps, be it noted, are not in the natural shell but in alabaster, so that the shell and stone lamps have, in a measure, joined forces. The forms adopted, however, were still based on shells of one kind or another and the same basis, though much simplified and conventionalized, can be traced in a number of other alabaster lamps, usually provided with a decorative device in the form of human or animal heads, found at Ur. The metal lamps of open type (gold, silver, and copper)[1] from Ur and Eshnunna are not quite so obviously shell forms perhaps, but the course of development followed in the derivatives from the natural shell enables one to see that these also are based on shells cut lengthwise.

The shells most commonly used as lamps are those of the pecten (oyster and scallop) and whelk groups. In the Orkneys and Shetlands lamps were made from suspended shells and were known as 'buckie' lamps—buckie being the Scots name for the *fusus antiguus* or red whelk. South Kensington Science Museum possesses examples of both oyster- and whelk-shell lamps. In the National Museum of Wales at Cardiff there is an example of each, both having been procured in Wales, and Mr. Peate wrote the author that the oyster-shell lamp was in general use in the Gower Peninsula (Glamorganshire) at least as late as the latter part of the nineteenth century of our era. Oil or fat was used as the fuel, and the example at Cardiff contains remnants of tallow, with a rag wick. Ireland and Cornwall, too, have used shells as lamps within recent times (as well as for candlesticks of an improvised nature). An old Cornishman told the author of such use within his memory, and the latter was just too late to rescue from the oblivion of the rubbish heap a tallow-filled shell which an antique-dealer friend had found in an old chest from Cornwall but had not recognized as a possible lamp or any article of utility at all. Two South Kensington examples of oyster-shell lamps come from the Duchy and another, of the coiled-shell variety, from Virginia, U.S.A. Was this latter the idea of some emigrant Scottish islander? It is dangerous to draw definite conclusions from the occurrence of shell lamps in particular areas as the idea is widespread and almost certainly has no common origin. In the Smithsonian Institution collection at Washington, for instance, there is a lamp made of one valve of a pecten shell of a type commonly used by the Aino of Japan, the fuel, as was the case in the Shetlands, &c., being fish oil, and the wick of fibre.

[1] The gold and silver examples are doubtless ceremonial burial objects only.

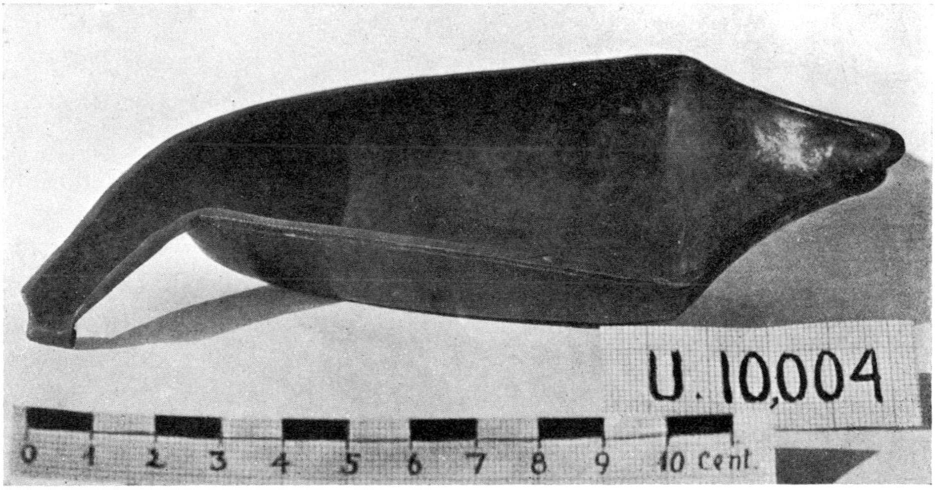

LAMPS FROM UR
(By permission of the British Museum)

So much for the direct evidence of the distribution of the shell lamp. For indirect evidence we have the unmistakable shell form of ancient pottery lamps of the Mediterranean area, which point plainly enough to the use of the shell itself as a precedent; generally speaking, the earlier in date the pottery lamps are the nearer they are to the actual shell in design. To this last fact there appears to be an important exception: lamps found at Tell Duweir (Palestine) by the Wellcome Expedition, and dating from the third millennium B.C., though closely related to those of shell form, are not only of finer pottery than that of later lamp finds but are almost square, four-spouted, and flat-based.[1] The flat base, in fact, as far as is ascertained at present in that area, appears to ante-date as well as follow the rounded base which is more characteristic of the true shell form. Later lamps from the same site, dated at about 1400 B.C., show an almost exact shell form, provided with a rim band, lip or flange, and have the rounded base of the scallop shell, side by side with a form similar in body but provided with a distinct foot such as is generally associated with a later development. (See illustration in the *Illustrated London News*, 16 Dec. 1933.) The only explanation which seems to offer itself is that the natural shell form was present at a much earlier date, though not, so far, found in such associations.

This form of pottery lamp based on the scallop shell is particularly associated with Phoenician influence. Its form indicates that it was originally evolved on the coast, though penetrating into the interior. At any rate, whoever invented it, the Phoenicians unmistakably spread it; it is found practically wherever they or the Carthaginians settled or traded. The Museum at Valetta, Malta, possesses a very large number of the shell-like pottery lamps found in excavations on the island and dated from about the middle of the second millennium onwards. Mr. Walters, in his *Catalogue of Greek and Roman Lamps in the British Museum*, lists sixteen lamps of what he calls the 'cocked-hat' type (for which perhaps 'shell-type' would be a better term). Of these, three have come from Cyprus, two from Carthage, nine from Sardinia, and two from Egypt. Nine are for two wicks, seven for one. The author himself has collected a dozen of the shell-form lamps, of which four came from Cyprus, one from Crete, one from Egypt, four (two-spouted) from Malta—the provenance of the others being unknown. The Maltese examples are commonly two-spouted as are most of the Sardinian examples in the British Museum. Cretan and Cypriote examples are generally one-spouted, Carthaginian running through a series of forms, single and double spouted, according to period.

Mr. Walters, in his introduction, refers to the true shell-type lamp as having been dated to as early as the twelfth to tenth centuries B.C., but finds at Tell-el-Hesy (Palestine) have been dated rather earlier than this, and recent finds at Tell Duweir considerably ante-date these periods; the lamps of Malta are

[1] 1937–8 finds include early lamps more of the saucer form than the shell-like lamps of the Middle Bronze Age following them.

sometimes said also to go back to about 1500 B.C. Cypriote examples are dated as from the eighth to the third century B.C., and Professor Petrie gives from the fourth to second centuries B.C. for Egyptian examples, though Phoenician contacts would suggest an earlier date. Some of the lamps in question appear to be wheel-turned.

To those who are convinced of the Phoenician contact with Cornwall, the use of the shell as a lamp there may appear to be another slight piece of evidence, but too much stress should not be laid on this in view of the widespread use of shells as lamps.

To turn again to Palestine, the lamps of the Canaanite period (second millennium B.C.) closely follow the lines of the Phoenician lamp and have obviously been influenced by the form of a sea-shell. Coinciding as they do with the early Phoenician lamps in both form and period, it seems that either they both have a common origin or that the Canaanite lamps were derived from the Phoenician. In the Early and Mid-Palestinian periods (particularly from Solomon's time onwards) the pottery lamps of Palestine follow a process of gradual development away from the pure shell form, by the acquisition of flat bases of more and more pronounced shapes, the transformation of the pinched lip into a definite wick nozzle, and the comparative deepening of the smaller body. Throughout the process, though, the shell origin is still traceable, especially if the development is followed step by step from its inception. The author has a large number of these Palestinian and similar lamps in his collection (one being of a folded edge type, often associated with Carthage, which suggests a different kind of shell); their precise provenance is in most cases unknown, but several appear to be modern, a fact which in this case means survival and not mere reproduction. Two were acquired in association with Cypriote ware and the presence of this developed form in Cyprus is confirmed by the ascription to that island of such a lamp in the Smithsonian collection, which also has a similar example from Syria. Five modern examples in the author's collection were said to have come from Corsica (and probably did). Two of similar shape from Cairo have an internal green glaze suggesting Arab medieval work.

The *Illustrated London News* in its issue of 29 November 1930 pictured a lamp of the Canaanite or Phoenician type from the excavations at Ras Shamra (North Syria). The lamps found there were not only in pottery but, unlike most of the shell derivatives, also in bronze. (The Phoenician connexion was here, too.) The Wellcome Expedition found another bronze example, of pure shell type, dating from about the fifth century B.C.

Between the Phoenician types and the earliest Greek lamps there is a gap, so far apparently unbridged, in the history of the lamp in Europe. Archaic Greece seems to have depended mainly on the torch for lighting, and when the first true Greek lamp appeared it owed little or nothing to the shell form. Early Egyptian lamps are known only from the hieroglyphics, a few wall paintings, and Herodotus, but were almost certainly of a simple saucer or bowl type.

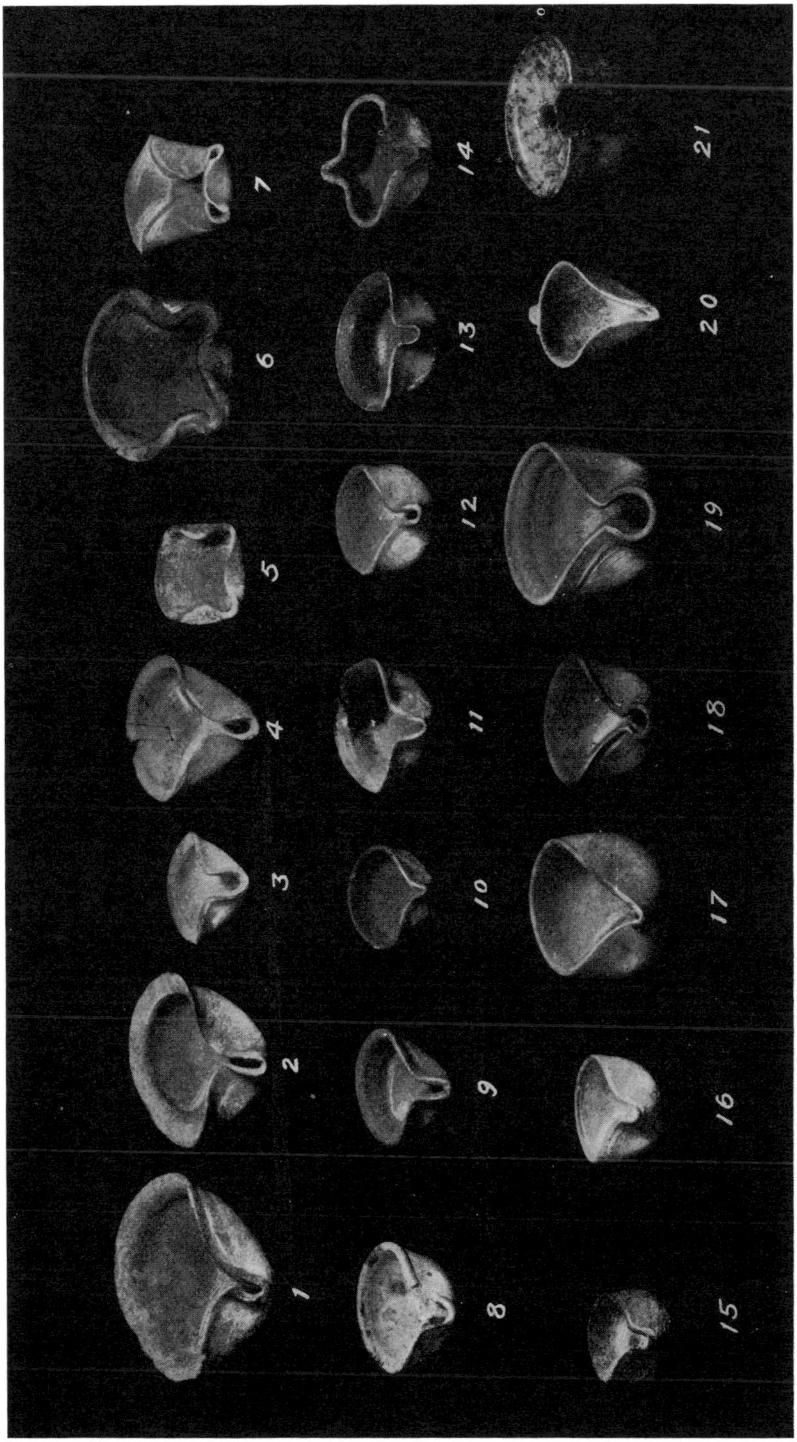

POTTERY LAMPS OF 'SHELL' FORM

In Palestine the shell-form pottery lamp not only connects with the Greek or classical type without a break in date but overlaps it as it is still in use, especially by the nomad Bedouin of Palestine and its borders, in whose tents the pinched lip, later form, burns, sometimes supported on a stand such as a branch cut to form a tripod. Somewhat curiously, until one remembers the Phoenician connexions, the later form was recently to be found in southern Italy, where it also gives the shape to the bowls of the glazed pottery pedestal lamps for olive oil. No doubt similar traditions are responsible for a variation, from the Sardinian tin-mines, in the author's collection and for a small leaden example of pure shell form on a flat tray, from Malta (cf. a Cypriote leaden lamp of shell form in the British Museum). The shell form is traceable, too, in certain blue and green glazed lamps of medieval Persia.

It is certain that the influence of the shell and even the use of the actual shell itself must have gone on continuously in many areas and that its use in modern times is a traditional one. It is extremely likely that the open iron lamps of the crusie type owe much in their inception and even in their form to the shell; this is the more probable in that the crusie and its relations were particularly prevalent in sea-lands, such as the Scottish and Irish islands, the Channel Islands, Cornwall, Brittany, Iceland, Norway, &c. In two of these quarters at least the actual shell was in use side by side with the 'crusie' and the Cornish 'chill'.

Neither Robinson Crusoe nor the Swiss Family Robinson are recorded as having thought of the idea! Primitive man was, in some respects, a more imaginative inventor than writers of fiction. Consciously or sub-consciously indeed his present-day remote descendants seem still to be constantly influenced by the inventions of the scientific dawn; gas and electric light fittings are frequently made in imitation of the shell!

[*For illustrations see Plates VII, VIII, and IX.*]

CHAPTER VIII

THE OPEN BOWL OR SAUCER LAMP

It is a curious thing that while, in most matters, burial customs and the climate have preserved for us so much material connected with the life of ancient Egypt that an almost complete picture of the latter is available (in many respects more complete than that of medieval life and certainly more complete than that of the life of the Anglo-Saxons of our own land), the one thing that seems shrouded in mysterious darkness is that of Egyptian methods of artificial lighting.

It is scarcely correct to say, as has sometimes been said, that nothing whatever is known of this; while few, if any, actual lamps have been identified, something is known of Egyptian means of illumination from hieroglyphs and tomb paintings. It may be more correct to say that the lamps of ancient Egypt are conspicuous by their absence from relics recovered by modern archaeological expeditions, but even this requires some cautious qualification, as it may well be that some of the lamps have survived but have not been identified as such. A study of some of the forms which the lamps appear to have taken strongly suggests the possibility of this being the case, as it is impossible to be dogmatic as to the precise use or uses of much of the pottery extant.

Herodotus, writing in the fifth century B.C., refers to a festival of lamps at Sais, during which the inhabitants all burned a multitude of lights in the open air around their houses. They used, he says, lamps in the form of flat saucers, filled with a mixture of castor oil and salt, on top of which the wick floated.[1] This is significant, as the flat saucer form, with a floating wick (and therefore not requiring a spout or blackening the edge), would be difficult to identify as a positive lamp without some sign or indication of its use. Neuburger, in *The Technical Arts of the Ancients*, suggests that the Egyptians might have used glass for their lamps. This is out of the question for the Pharonic period, but it is of interest to note that the earliest examples of glass lamps are of floating-wick type and were found in the Fayoum. It is extremely probable, in fact, that the floating-wick lamp *in toto*, with its particular development in glass forms, originated in Egypt, as it is certainly not Greek or Roman in form and seems hardly likely to have appeared suddenly, without antecedents, in the Byzantine period, in which it first becomes prominent. Neuburger also refers to clay bowls about 3 inches high 'similar to present day nightlights and showing no provision for wicks' being found on two sites in Egypt. This again suggests that some of the pottery in saucer forms or as shallow bowls, found in Egypt, may well be the lamps.

[1] Diodorus, Strabo, and Pliny also mention the use of castor oil by the Egyptians, and Mr. Lucas, in *Ancient Egyptian Materials*, suggests that linseed oil was also used, as it is still by the poorer classes in Egypt.

A further reference by Neuburger relates to limestone stands about 3 feet high used on the occasion of festival illuminations to support flat granite bowls with no special device for receiving wicks. While he does not give chapter and verse to enable these stands to be identified by others the allusion connects both with Romano-Egyptian finds in the Bucheum (in pottery) and with certain stands depicted in tomb paintings to be mentioned hereafter. The most direct evidence as to a lamp in general use at an early stage is perhaps provided by the hieroglyph 𓊸. This, though usually interpreted as a bowl for incense (*Gard. Egn. Gr.*, p. 489, R. 7), certainly suggests a simple bowl with a floating wick in the centre and without any form of spout or nozzle. Confirmation of the form occurs in the wall paintings reproduced on Plate VII of *The Rock Tombs of El Amarna*, Part III, 'Tomb of Huya'. In this illustration, too, the lamps are shown on stands. Each of the lamps shown is flanked by vessels described in the letterpress as for wine 'set so that the servants can see to fill the cups'—but might not these have been oil vessels for refuelling?

The lamps just referred to are with flat bases and straight splayed sides as shown in the hieroglyph, but other forms occur elsewhere. In a sacrifice scene (Plate XV, *Rock Tombs of El Amarna*, Part IV), a round-bottomed bowl with a rounded rim is shown, bearing three flames. A similar form of vessel is shown as a brazier; this is distinguishable by the fact that embers are shown above the rim, but it suggests that such bowls were not confined to a single purpose and tends to underline the suggestion the author has made as to the difficulty of definite identification. Still a third form is a plain round-bottomed bowl without the rounded rim; this is shown both with single and triple flames in a picture of the adoration of the Sun (*Tombs of El Amarna*, Plate VIII, Part II), and similar lamps appear in sacrificial scenes elsewhere, held in what appears to be a modelled hand at the end of an 'arm' held in the hand of the worshipper. These, of course, may be, in such a case, incense burners rather than lamps, but it looks as if the same vessel was used for both.

In the *Journal of Egyptian Archaeology*, April 1924, vol. x, part 1, N. de Garis Davis has an interesting and detailed article on 'A New Kingdom Lamp'. The particular device to which he primarily refers would appear to be more in the nature of a torch, consisting as it does of a pyramid (in one case) or a cone-shaped object, probably of wax, decorated with red and yellow bands and set on a short pole stuck in the ground. These are sometimes shown flanked by burning tapers, each formed of three white strands twisted like a rope and lashed in the middle and at the ends with red twine. Another shown with the article, however, is a cup-shaped bowl on a columnar stand, from which there rises either a long vertical flame or some form of candle or torch. A pair of 'lamps' shown in the papyrus of Ani (British Museum) present a somewhat similar form, but the base of the column is more widely splayed and the bowl at the top is a wide-mouthed splayed form instead of a cup. In this representation the projection from the centre of the bowl seems more clearly to be a flame, as it comes up in a wavy curve, right from the bowl. The top portion of the 'flame' is red but the bottom is white, a fact which should carry some significance in relation to the kind of fuel used. The so-called 'candlesticks' from Tut-ankh-amen's tomb, containing a thick length of rope-like wax in a handled cup and requiring replenishing with oil or fat, suggest some sort of a cross between a lamp and a torch-holder.

Whether the 'stand' arrangements described are to be more properly classified as lamps or torch-bearers, they present an interesting comparison with red pottery columns surmounted by shallow *spouted* bowls, found in the excavations at the Bucheum by Mr. Oliver Myers of the Egypt Exploration Society. Side by side with these (dating from the Roman occupation period) are two other forms—one a shallow spouted saucer on a low foot and the other a cup form, with splayed base, also spouted. While the presence of a spout indicates a change in method suggested by Greek or other influence (since the lamps of the native dynasty epochs seem to have all been of the floating-wick type), it may, on the other hand, be a perfectly natural improvement, and the fact that the Bucheum open lamp is considerably more primitive than the contemporary Roman lamps points to the survival of a native form of an earlier date. It is interesting and curious to note that this open saucer form of lamp is exactly the same (except as to usual size) as the little Indian 'chirag' or earthenware lamp which has survived to the present day though in danger of rapid extinction through the rivalry of the European hurricane lamp—and the converted cigarette tin! The chirag, too, was used in large numbers for festival illumination, in precisely the same manner as in the Sais festival.

That Egypt had more elaborate fittings for lighting than the simple bowls used alone is shown by the discovery made by Mr. Howard Carter's expedition that some of the alabaster vases of Tut-ankh-amen's tomb were really palace 'lamps'. This was indicated when a light placed inside a vase which appeared to be of plain alabaster revealed coloured picturing and demonstrated that the bowl was double, with the picture on the outside of the inner bowl (and therefore only visible when illuminated from the inside). Strictly speaking, however, these vessels were, in effect, lamp 'shades' or holders and the lamp itself was again probably the simple float-wick bowl or saucer.

The simple bowl lamp is one of the many points of similarity between ancient Egypt and ancient China. While the candle seems to have been, in more recent centuries at least, a commoner form of illumination than the lamp in the Far East, plain open lamps have been in use there from a remote period, and the Royal Ontario Museum at Toronto, Canada, has a collection of very early types, from a limestone lamp or lamp-stand (a hollowed-out cup on a low foot) of the Shang-yin dynasty (1766–1122 B.C.) onwards. Notable and typical examples in this collection are a number of bronze shallow saucer lamps on pedestals or low feet, belonging to the Late Chou dynasty (third or fourth

(By the courtesy of the Royal Ontario Museum)

(*a*) ANCIENT CHINESE LAMPS, ROYAL ONTARIO MUSEUM, TORONTO

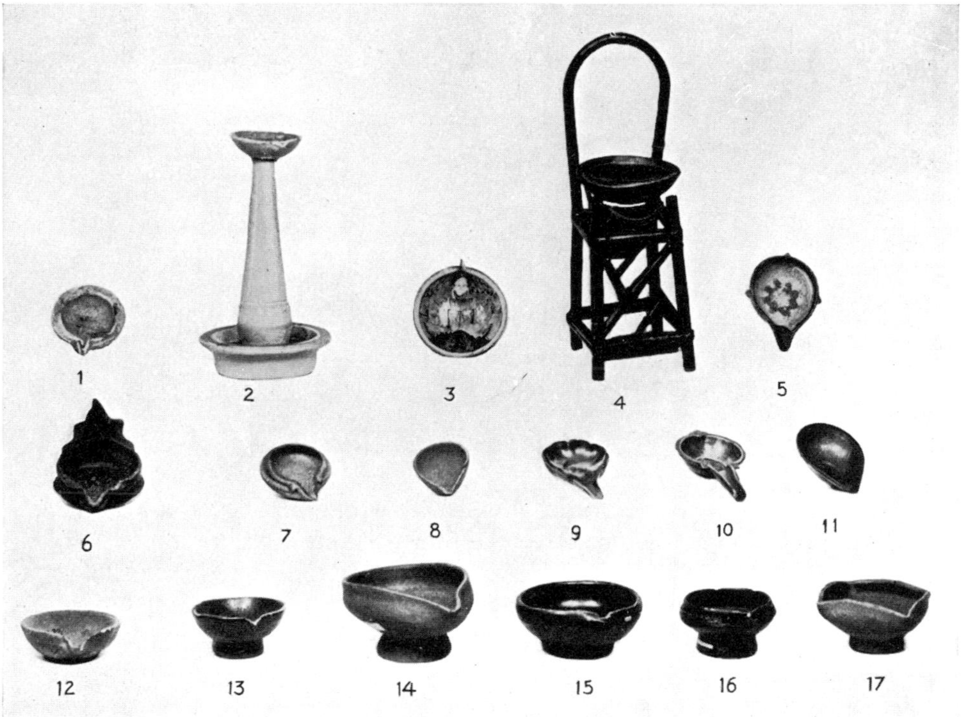

(*b*) SAUCER TYPE LAMPS

century B.C.?), which present an interesting development in the form of a low spike apparently to act as a central wick support such as sometimes occurs in the medieval cresset lamps of Europe. At a later date this wick support disappears and the lamp reverts to a simple plain saucer. Sometimes ancient Chinese lamps were on elaborate supports, as a bird standing on a tortoise and holding in its beak the stem of a circular shallow oil bowl. In the first year of Ho-p'ing (28 B.C.) an official made for the 'Inner apartment' a Feng Huang or phoenix lamp. (Both the tortoise and the phoenix were symbolical—the former of longevity and the latter of a felicitous reign.) The *Illustrated London News* for 11 January 1936 illustrated an array of Chinese bronze 'saucer' lamps on an emblematic 'tree' (from the Toronto collection) some two thousand years old, together with grey pottery lamps of similar simple form on a stem and mythological base, dated as second to fifth century A.D. Later Chinese lamps mostly present the form of simple round bowls or saucers, in pottery, bronze, or iron, sometimes with a back projection rather reminiscent of the leaf and other appendages of some Roman lamps.

The Smithsonian Institution collection includes saucer lamps from Korea, China, and Japan, in pottery, porcelain, and brass, and quotes the use of pewter and iron also; these lamps were usually set on stands of porcelain, metal, or bamboo. Occasionally, the saucer form is adapted as a hanging lamp; an example from China was pictured in the *Illustrated London News* and another from Japan is in the Smithsonian collection. In the South Kensington Science Museum there are several shallow saucer lamps from China in iron and pewter.

Excavations in Northern Annam (*Illustrated London News*, 23 Dec. 37) yielded the saucer-form lamp on a pedestal or stand with splayed (not dished) base.

Similar saucer lamps to those from China occur, in various materials, in Polynesia; coco-nut bowls with float wicks in New Guinea are paralleled by the use of the same thing in India (sometimes floated down the Ganges for the purpose of divination). Shallow bowls or saucer lamps have been used in the Shan States and this simplest of forms of the lamp seems to be especially prevalent from Burma eastwards—and eastwards must be taken to include crossing the Pacific if examples from Alaska and New Mexico are to be regarded as indigenous. Or is this evidence for the thesis of the 'Asia to America' culture protagonists? ('Lamparistas'—saucer lamps for floating wicks in peanut oil—are, the author is told, still used in Brazil, but these are probably of Portuguese origin.)

Occasional examples of spoutless saucer lamps are found in India, though the spouted form is more normal there. Three iron lamps reached the author from Malabar which were plain and shallow round saucers, suspended by a hinge from a 'pot-hook' stem or rod which was inserted into a heavy wood block base (Plate XVII, *b*).

Isolated examples of plain saucer or shallow bowl lamps crop up elsewhere, as, for instance, pottery cups showing signs of fire found on a Saxon site near Winchester in 1934 (now in Winchester Museum); the bases of these particular specimens, however, appear to have been broken, and they were probably stand or pedestal lamps similar to one from the same area in the author's possession, which suggests the ancestry of the later open but spouted saucer stand lamp in earthenware found in north-west and central Europe.

The idea of the use of a saucer or bowl as a primitive oil lamp is of course a fairly obvious one and may well have originated independently of the stone lamp and the shell, by the adaptation of ordinary domestic pottery utensils for the purpose.

The spouted saucer lamp, though technically in a different class, is a natural development from the plain type, especially where the wick was allowed to hang over the side instead of floating in the middle; in such a case it called plainly for some provision of a resting place, and even some primitive stone lamps show a groove in which to lay it. The greater plasticity of clay invites the definite formation of a wick channel in the earthenware lamp even where it had not been suggested by the outline of a shell. As has already been noted, the lip or spout is present in open lamps from ancient Egypt and modern India; it is also present in Minoan lamps which do not appear to be directly derived from the shell form and which may possibly betray Egyptian influence (just as the Phoenician 'shell' form appears, comparatively late, in Egypt).

The spouted saucer lamp is widespread through the Middle and Near East, plain or decorated and in pottery or metal (usually bronze or brass), and dates from an unknown beginning to the present day, generally, as in the Indian 'chirag', in the form of a round or oval bowl with flat base or low foot, with a simple lip for the wick and no means of keeping the latter in place. In Layard's *Nineveh and Babylon* there is a description of an Arab lamp (mid-nineteenth century) with four burners—a simple dish with four lips for the wicks, supported on a sharp iron rod driven into the ground. Does this owe anything to the open lamps of Luristan? A spouted saucer lamp in terracotta ware with an internal brown glaze and a low foot was procured by the author from the sale room of the Arab Museum at Cairo.

Africa, outside the romanized or orientalized North, had few lamps. Nigeria, however, has had and still has pottery lamps of the lipped saucer type with low foot—strongly reminiscent of the above-mentioned lamp from Cairo and undoubtedly due to Moslem influence; they are rapidly disappearing in the face of European importations and native improvisations from old tins, as is the case with the not dissimilar Indian peasant lamp. Usually, the Nigerian lamps are simple lipped bowls with one spout, but one type has a burner at each of four corners; Miss Deakin, of the Achimota Museum, to whom the author is indebted for a specimen of this type and for other lamps, states that these were not lit together, but as one lip was broken another was put into use.

Other simple lipped open bowl lamps have existed until quite recent times in European countries (particularly in the northern and Mediterranean areas), but these have taken a pedestal form of their own which brings them into a peculiar group dealt with elsewhere.

[*See Plates X, XVII, and frontispiece.*]

CHAPTER IX

THE GREEK LAMP

WHERE Greek and Phoenician influences have mingled, the shell-form lamp has been found on Greek sites, but the Greek lamp proper seems to have little, if any, connexion with the shell and probably developed from a simple saucer or bowl. It was a late introduction so far as Hellenic Greece was concerned, Greek artificial lighting before the seventh century B.C. being a matter of torch or pine splinter and brazier only, according to the dating of lamps found.

A number of lamps from Crete, dating from the late Minoan period (before 1000 B.C.), show no direct derivation from the shell type (widespread at that time) but are of simple open type, though sufficiently developed to have a distinct and serviceable nozzle. Pfuhl states that the Minoan lamp was derived from Egypt, and Clement of Alexandria, doubtless with earlier tradition in mind, ascribes the invention of lamps to the Egyptians. It is sometimes objected to this that no definite lamps have come from Egyptian excavations, but if (as is suggested, at a later date than the dynastic periods, by Herodotus) these were simple saucers, it is more a matter of failure to recognize them than of their absence. Assuming tradition to be correct (as it more often turns out to be than some scientists are willing to allow), the Greek lamp would have been derived from a different source from the Phoenician shell-type lamp and was an alien importation, either from Egypt direct or from Egypt via Crete. (It is a curious point, by the way, that our very word 'lamp' is derived from the Greek word, 'lampas', for a torch!)

An additional point in favour of the Egyptian origin is scored if a certain Corinthian vessel of the seventh and sixth centuries B.C., which Mr. Walters (*Catalogue of Greek and Roman Lamps in the British Museum*) suggests was the Corinthian form of lamp, is taken into consideration, as this has no spout or nozzle for a wick and must, if used as a lamp, have had a floating wick such as is indicated by some of the Egyptian wall paintings and by Herodotus's allusion to the Egyptian lamps of his time.

The Greek lamp proper commenced its history in the sixth century B.C., and from that period the scientific Greek mind got to work on its improvement from its primitive beginnings. The first stage was to recurve the rim, to prevent the spilling which would have taken place from the ordinary saucer. (Here the wisdom of the ancients has lain neglected: the author only calls to mind one railway company in this country which realizes the advantage of turning in the rims of the teacups used on its restaurant cars!) Even at this early stage (and earlier, in the Minoan examples) a handle appears—only to disappear for a time in most of the later Greek lamps.

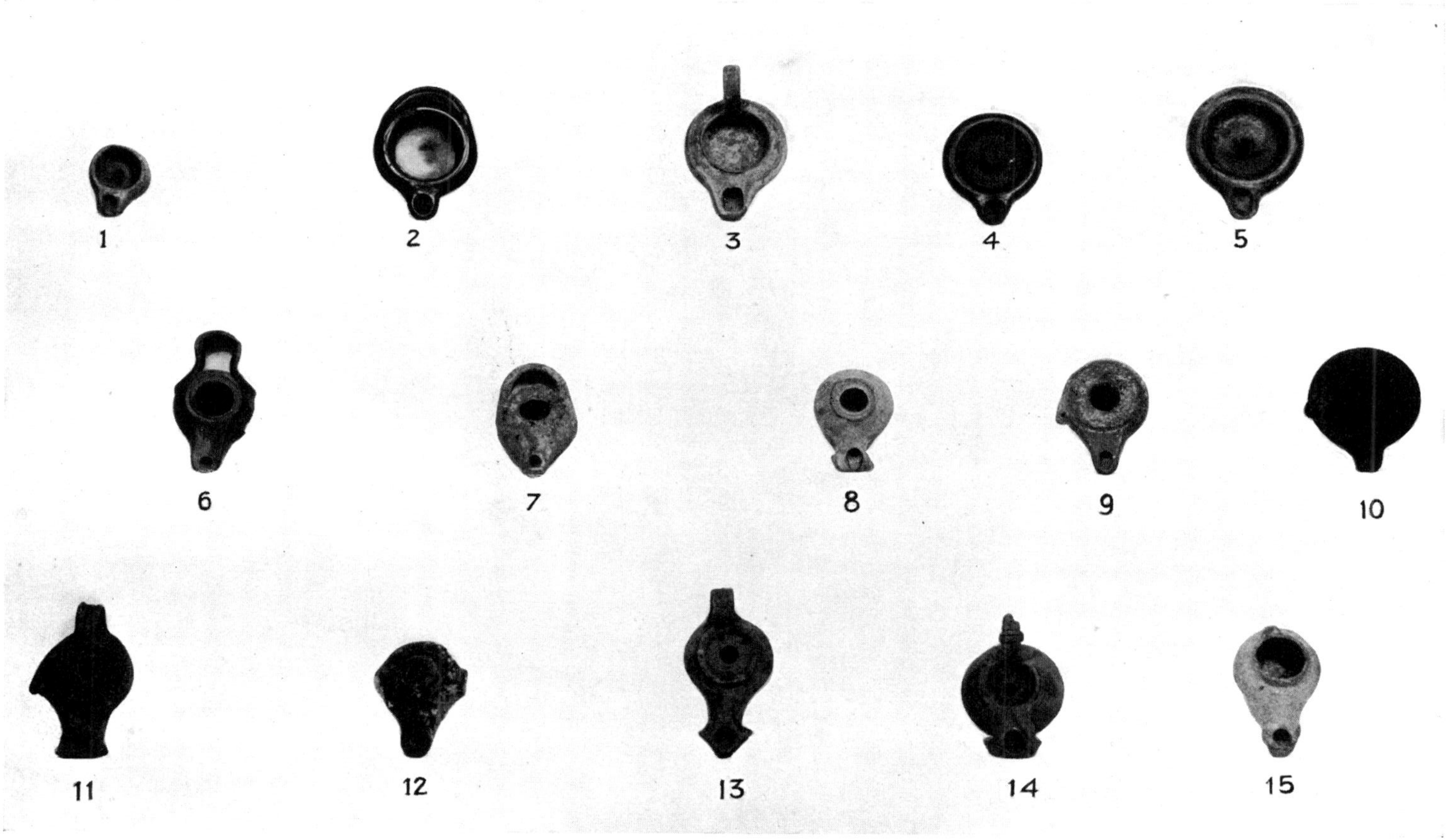

ANCIENT GREEK POTTERY LAMPS

Lamps of the Phoenician type had largely been made by hand. The Greek lamps from the sixth to the third century B.C. were made on the potter's wheel; hence the absence of decoration, which would have had to have been subsequently applied and was no doubt regarded as unnecessary for such a purely utilitarian object. The lamps of this period were usually coated with varnish or glaze, the earlier forms mostly in the typical black of Greek pottery.

Right at the beginning of the development of the Greek lamp, another improvement, absent in its prototypes, appears. The nozzle is bridged, thus providing a means of holding the wick. These two developments—the recurved rim and the bridged nozzle—start the closing-in process and pave the way to the totally enclosed lamp of the late Greek and Roman periods. It is a curious example of the ebb and flow of progress that the comparatively recent open lamps of the Mediterranean (particularly those of southern Italy and Sicily) and the simple crusie types of western and northern Europe are devoid of such useful devices.

The shallow open saucer lamp with the bridged nozzle, with or without a horizontal handle and with a recurved rim gradually widening, extends from the sixth century into the fourth century B.C. (British Museum forms 15 to 26) with variations of detail. It has been found throughout the Greek areas, with examples in Cyprus and Sardinia. Broneer (*Terracotta Lamps of Corinth*) dates such lamps without raised bases but well developed before 500 B.C. Bases appear at the close of the sixth century B.C.—high bases on lamps of the fifth to fourth century B.C. becoming less common later. Bases are flat at first, rising in the centre (to concave form) later. Some of the sixth century B.C. lamps are unglazed, but many are glazed; according to Broneer, those of Corinth are mostly in yellow clay.

The next stage from the open saucer types is what Mr. Walters calls the 'Attic' type of lamp. In this group, which covers a number of different forms (B.M. forms 34 to 41 inclusive), the chief characteristics were the deeper body and the pronounced nozzle, the latter often projecting considerably from the body of the lamp. Also, the closing-in process went on until, in forms 36, 38, and 39, the third-century lamps (in which the opening became merely a large filling hole) were foreshadowed. These lamps, dating from the fifth to the third century B.C., come under Broneer's Types VI and VII, his first five types (sixth to the fifth century B.C.) coinciding with the British Museum saucer forms, though the methods of grouping differ. From the end of the sixth century B.C. to the fourth century B.C., practically all Greek lamps were glazed outside (Broneer), but Attic glaze deteriorated through the fourth century.

Overlapping the Attic types and the lamps of the third century B.C. (though Broneer seems to put some earlier) came a group of open saucer types provided with a central socket (forms 27 to 34), the idea of which is said to have been either for suspending the lamp or for placing it on an upright spiked support.

These groups complete the range of lamps which Broneer classes as 'Greek' lamps, a marked difference in development in lamps of the third century B.C. onwards, drawing a line of distinction between the former and what he calls 'Hellenistic' lamps.

The later Greek or 'Hellenistic' lamps of the third century B.C. (B.M. forms 42 to 49) are generally more or less globular in form, deep-bodied, long (and usually blunt) nozzled and frequently with a fin or projection at the side (often pierced and probably originally intended for the insertion of a 'pin' for raising the wick). They are often 'unpainted' but with a fine slip of clay outside. All through the 'Greek' period there had been a tendency to decrease the width and increase the depth of the lamps, and this tendency culminates in the third-century groups. Wick holes are round or spoon-shaped.[1] The earlier lamps of the type have flat tops, but later they become more markedly globular in form; raised bases are taken by Broneer as a sign of earlier work than those which are flat. Large numbers of the third century B.C. types have been found in Egypt, in coarse wares, and the lamps in general are made of coarser pottery than the earlier types and are frequently unglazed. It is in this period (the third century B.C.) that the handle most completely disappears, for some reason one cannot fathom.

Up to this point Greek lamps had been wheel-made, the nozzles being stuck on by hand. In the second century B.C. the method of manufacture changed to moulding (appearing in Asia Minor at the beginning of the century but not common in the west until later); decoration in relief then became a comparatively easy matter, the decorations being mostly of a conventional nature, such as rays, herring-bone patterns, palmettes, &c. In the 'Knidos' group of lamps (Broneer XIII, B.M. forms 56 and 57), the lamps themselves (in dark grey clay, unglazed) are modelled, with designs (principally leaves and flowers) impressed from moulds. Their nozzles are 'splayed'. Other groups of decorated lamps of the second and first centuries B.C. are peculiar to Aegean and Asiatic sites, and in the first century B.C. a type of lamp known as the 'Delphiniform' (from the dolphin-like fin on one side) appears (especially in Egypt and North Africa). These last are usually undecorated save for simple patterns around the central filling hole and have straight-ended concave-sided nozzles; they are obvious developments from the third century B.C. types.

Greek nozzles in general differ from those of the Roman lamps in that they are spoon shaped (early lamps), blunt-ended (delphiniform types), splayed (Knidos and a few delphiniform lamps), ending in defined rimmed circles (many lamps of B.M. 52–3 and some of B.M. 55) or lozenge shaped (as in B.M. 54, Ephesus type). These features persisted in some Egyptian lamps which may be of later date but which were clearly influenced more by Greek than by Roman traditions.

In the Greek lamps, throughout, the rounded base of early 'shell' lamps was

[1] Sometimes these lamps have two or even three nozzles—usually one only.

replaced by a flat one in the saucer lamps, and, as already indicated, this was accentuated until, in the lamps of the later Greek period it became a low foot (which, in its turn, is ousted by a reversal to the flat base in most Roman lamps). The periods immediately preceding the Christian era, though retaining the convex top to the pottery lamp, foreshadowed the depressed (concave) bowl of the Roman lamps by the presence of a depression immediately around the filling hole. The Knidos lamps, too, with their handles in the form of two 'ribs' banded together and other late Greek lamps with ridged handles, suggest possible origins for the almost universal grooves of the ring handles of handled Roman forms. Broneer suggests that the handles on the Knidos lamps were in imitation of metal lamps having handles of double rods held together by small hoops.

Makers' names are absent in Greek lamps of the pre-Roman era but some of those of the first century B.C. (particularly those from Egypt)[1] carry a single letter or monogram on the base to denote the maker or the factory where they were produced.

Olive oil was the principal fuel used, as in the case of most primitive lamps of the Mediterranean area right down to modern times.

Somewhat analogous to Greek lamps, and possibly influenced by them, are the Parthian lamps of 200 B.C. to A.D. 200, a large and representative assortment of which were unearthed at Seleucia. The line of development was similar in that a few examples of open saucer lamps with bridged nozzles were found and one open lamp with unbridged nozzle and an open bowl, on a stand or foot, but the majority were small lamps with circular bodies, large apertures and long projecting nozzles rising at an angle from the base of the reservoir. They were mostly made on the wheel, with the nozzles stuck on separately (and noticeably) and generally glazed inside and out with a glaze ranging from white or grey through greenish-grey to green, in some cases the nozzles being tipped with a darker blue-green, which it is suggested may have been especially resistant to the heat of the burning wicks. A curious variation in form was a lamp with a nozzle projecting *downwards*, which could only have been used on the edge of a shelf or ledge. Only a few of the glazed lamps were decorated; some of the unglazed, however, were elaborately ornamented, the decoration consisting of rays, floral designs, &c., much after the style of the later Greek lamps. Handles were usually little more than pierced lugs at the back of the reservoir, though here and there the ring handle appeared and sometimes a solid handle in the form of a human head or torso. The decorated forms, as in the case of late Greek lamps, were moulded longitudinally—the base being one piece and the upper portion another. Multiple lamps were present but uncommon.

[1] There was a 'local' Egyptian manufacture of Greek lamps, forms 42 to 50 and Graeco-Egyptian forms (q.v.).

Mr. N. C. Debevoise (U.S.A.), who has illustrated a selection of these lamps in *Parthian Pottery from Seleucia*, suggests that they were usually used with sesame oil, on account of the expense of importing olive oil. This is probable, as in all countries the most readily available oil or fat was the illuminating fuel in early ages.

A somewhat similar lamp to the Parthian types in Birmingham Museum, is ascribed to 'Nineveh, period V', and a number of tea-pot-like pottery lamps of Assyrian origin in the same collection have upturned nozzles. Probably, therefore, the Parthian lamp was a hybrid between Greek and Assyrian types.

The foregoing applies, of course, more particularly to the history of the pottery lamp. Bronze lamps of both the shell and saucer types have been found—a lamp from Luristan, Persia (*c.* 400 B.C.), has a boat-shaped reservoir on a pillar pedestal, with a decoration of an ibex and lions. Greek bronze lamps, however, though not unknown, are scarce, and generally follow more or less the pottery forms (with exceptions, which include a British Museum example of a fourth-century B.C. lamp in the form of a greyhound's head).

GREEK LAMPS, IN POTTERY, BRITISH MUSEUM CLASSIFICATION

(1) Open saucer-shaped forms. Form 15, &c.
(2) Small plain open bowls. Forms 16 to 20.
(3) Flat open type, without handle. Forms 21 to 24 (form 24, two nozzled).
(4) Flat open type with handle. Forms 25 and 26.
(5) Flat open type with central socket. Forms 27 to 33.
(6) Deep bowl, with central socket; no handle. Forms 34 and 35.
(7) Deep bowl, no handle or central socket. Form 36.
(8) Deep bowl with handle, forms 37 to 41.
(9) Body of double convex section, with small orifice, no handle. Forms 42 and 43.
(10) Deep body with projection. Forms 44 and 45.
(11) Body of double convex section with lateral projection. Forms 46 to 48.
(12) Delphiniform type, with pointed nozzle. Form 49.
(13) Delphiniform type with blunt nozzle. Forms 50 and 51.
(14) Body-watch shaped, with double projections or ear-handles. Forms 52 and 53.
(15) 'Ephesus' type. Forms 54 and 55.
(16) 'Knidos' type. Forms 56 and 57.

[*Illustrations on Plate XI.*]

CHAPTER X

THE ROMAN LAMP

THE true Roman lamp, though differing substantially from Greek forms, is undoubtedly descended from them. The oldest Roman lamps (e.g. those of the third century B.C. found on the Esquiline) are more Greek than Roman in type and lamps from south Italian sites show a good deal of Greek influence. Lamps of the second and first centuries B.C. found at Carthage and elsewhere show transitional forms.

Like the lamps of Greece, those of Rome were frequently used, not merely for domestic lighting but also in religious ceremonies, as votive offerings and sometimes as funeral deposits, so that a Roman lamp need never be rejected as spurious because it shows little or no signs of burning. Caligula had theatrical representations and Domitian gladiatorial combats by the light of numbers of the ordinary lamps; it has been suggested that such uses were the origin of the gladiatorial designs on many of the lamps found, but, no doubt, devotees of the games would also have chosen such designs for their domestic lamps in much the same way as the football fan of to-day collects photographs of his favourite teams and players—wherein lies a suggestion, to electric light and gas undertakings, of a means of increasing their sales of shades and the like!

Two factors, in particular, led to the marked differentiation between Greek and Roman lamps. One was the closing-in process, which had become complete in the last century or two before the Christian era, and the other was the change in the method of manufacture to making in moulds (the top and bottom halves separately and subsequently joined together), a method invariably employed for Roman pottery lamps. The main point of divergence from the Greek lamp of the later period was the substitution of the concave top for the flat or convex form; the typical Roman lamp, circular in form, has a depressed bowl which would have facilitated filling without spilling and one or more small filling holes in the bowl (these being closed by a wooden plug when not in use). The nozzle or spout had more or less projection from the body but seldom a long one in proportion; where a handle was present, this was usually a grooved 'ring' handle, placed, except in the case of certain unusual forms, vertically, in a line with the axis of the lamp. The wick of vegetable fibre (e.g. from the plant verbascum or sometimes of tow or papyrus) was trimmed with a pair of tweezers and must, judging from the author's own experiments, have required constant attention to keep the flame going. The common single nozzle lamp was carried by hand or stood in niches in the walls or on brackets or stands (wrongly called 'candelabra'). Larger lamps were sometimes suspended by chains.

The form of the Roman pottery lamp and the method of manufacture by moulding lent themselves admirably to the development of decoration, already commenced in the moulded or hand-modelled lamps of the later Greek period. In addition to such more or less conventional ornamentation of the rims or sides of the lamps, the Roman potters commenced to utilize the bowl or *discus* as presenting a surface positively inviting design. The range of design through the first three centuries after Christ is tremendous, covering the classical deities, mythology and legend, animals, foliage, historical, and literary subjects, the games, and, most interesting of all, scenes from the ordinary social life of the period. Works on Roman life frequently draw on scenes depicted on the lamps for their illustrations and, if the play on words may be forgiven, these lamps may be said to have thrown light on the domestic and social life of the ancient Romans in both senses. It is for this reason that the pottery lamps of the Roman period are of particular interest to the author and of special value to others who are not primarily concerned with the history of lamps or lighting.

Although decorative design had appeared on lamps from North Africa in the second century B.C., it was not until the Roman imperial period that actual subjects, as distinct from conventional decorative designs, appear. Roman potters understood the art of effect as well as the most intelligent modern advertisement writers and their designs are never overcrowded, the number of figures depicted seldom exceeding three except in abnormally large lamps. There is at least one example, however, which depicts a circus with spectators, though the 'gate' would have been far from satisfying to a modern football manager—or even a county cricket committee for that matter!

Amongst the later lamps the deities depicted include, in addition to the classical divinities themselves, Isis and other Egyptian gods and goddesses. Amongst the purely classical figures Zeus and Eros (Cupid) are the most common—the latter in various aspects (even with the lamp's predecessor, the torch). The lamp-makers seem to have made the same mistake as poor Psyche, since Cupid has always preferred the darkness! Next to Zeus and Eros, favourite legendary or symbolical subjects were such as Victory (common), Fortune and Hercules (very common indeed). The gladiatorial subjects include, not only single gladiators and combats, but the details of gladiatorial armour detached from their wearers or users. Animal designs include lions (probably the most common), panthers, bears, boars, deer (also very common), horses (not usually alone), oxen, sheep, goats, dogs (very common), rabbits or hares, eagles (common), the peacock, cocks and hens, crocodiles, fish, &c.—sometimes as individuals but often in combination, as a lion attacking a deer or a dog chasing a hare. The dolphin is commonest on Mediterranean sites, the dog in the East. The donkey appears turning a mill on two examples in the British Museum; both are from London but the mills are identical with those extant in Pompeii.

In addition to the designs on the bowls or disci, Roman lamps frequently carry (generally on the base) inscriptions in relief (formed in the moulding), stamped (with a bronze stamp) or engraved with a pointed instrument. Inscriptions in relief are generally on the group consisting of B.M. forms 90 to 94

PLATE XII

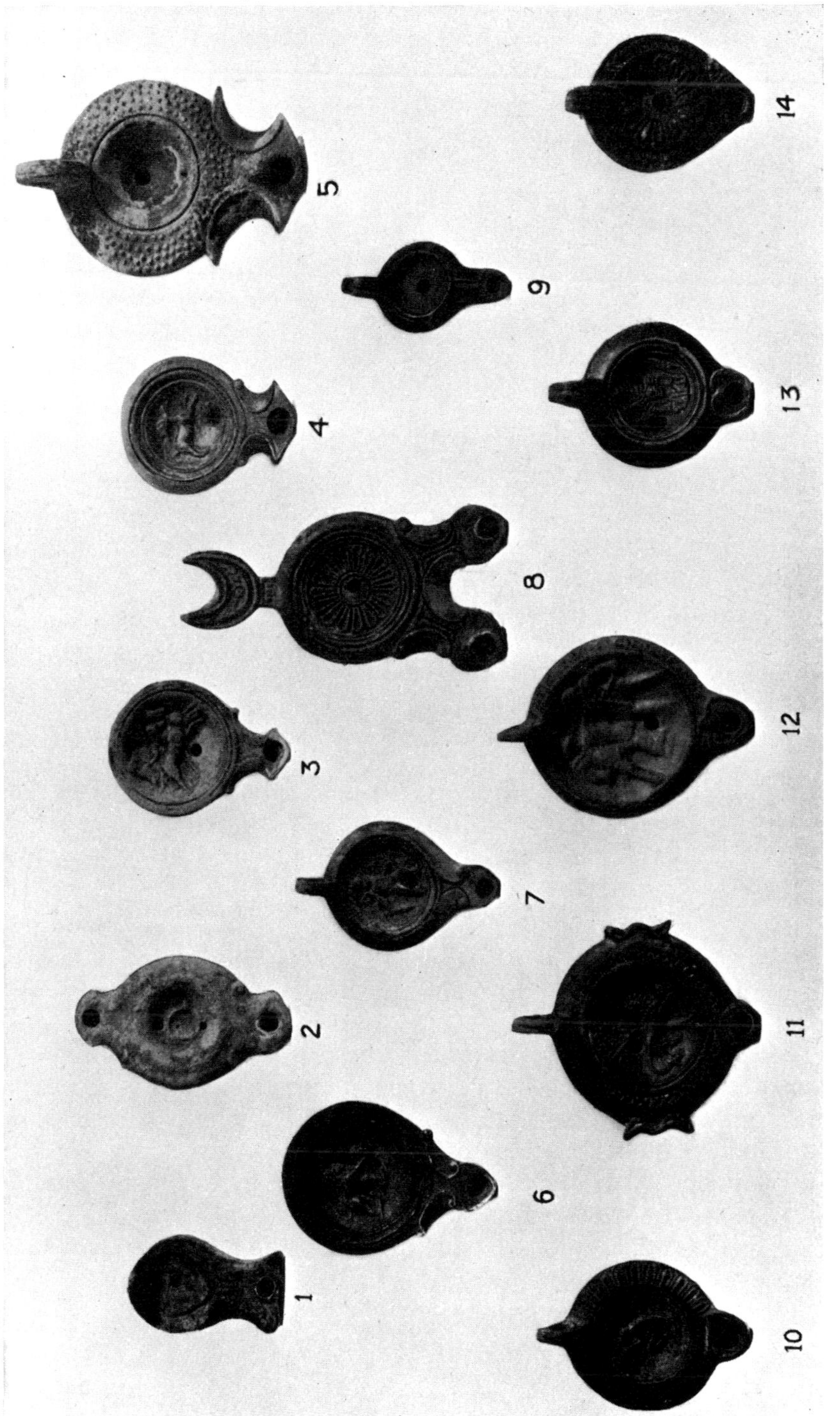

ANCIENT ROMAN POTTERY LAMPS

(Fink's Class III). Inscriptions include dedications (on lamps given as presents or commemorating some particular event as in the case of 'New Year' lamps), subject descriptions, and formulae addressed to the public, but the most common inscriptions are those denoting the maker or factory by whom or at which the lamp has been produced—sometimes a simple letter or emblem serving as a trade-mark. Potters' names are more common in the second-century lamps than at any other period and are usually in the form of three names—praenomen, nomen, and cognomen, abbreviated, stamped hollow or incised. Fink's Class III lamps are again exceptional in having a single name in the genitive. Late examples of Egyptian origin show a constant recurrence of the 'trade-mark' forms, the letter 'A' and the palm leaf being particularly frequent.

The most commonly used classification of Roman lamps is that of Herr Fink (of Munich) based on the form of the nozzle and grouping true Roman lamps into four classes only. Class I covers lamps with rounded nozzles flanked by volutes and handled or otherwise (= Broneer XXIV). Class II contains lamps with nozzles having broader angular terminations flanked by volutes, usually without handles, but there seems little doubt that these numberings should be reversed, since the splayed or angular nozzle betrays a Greek influence which puts it earlier than the plainer type; Broneer, who includes these in his form XXIII, dates those with base rings as earlier than those without. Herr Fink's order of placing is based on the idea that Class II shows more development (a moot point, having regard to the greater prevalence of handles in Class I) and decoration consisting of more purely Roman subjects. Both these classes belong to the first century A.D. and more particularly the earlier portion of it. The British Museum divides these two classes into twelve forms and places those with the angular nozzles as earlier; Loeschke agrees and gives the latter as dying out in the reign of Tiberius. Fink's Class I includes an interesting group of lamps having attachments to the handles, such as leaves, crescents (symbolic of Diana), &c., often two-nozzled. These are said to have been particularly prevalent among lamps from Roman Egypt, the home of many variations of the Roman lamp. Fink's Class III (Loeschke types IX and X, B.M. forms 90–4)—sometimes referred to as 'factory lamps'—are a complete break-away from the general development of design and are alleged to have been copied from bronze lamps. They are without decoration (except for comic masks in some cases) and have a sunk centre around which is a raised rim, a shallow groove passing to a rather long nozzle. South of the Alps they were usually unhandled, but a variety common in Britain (sometimes made there) and north of the Alps is often handled. Most of these lamps have the maker's name on the base in raised letters, as mentioned above. They are generally dated as belonging to the last quarter of the first century A.D. and the first quarter of the second, but specimens found at Caerleon and now in the Cardiff Museum are given as having been made late in the second century A.D. at Holt, Denbighshire, by one Sextus.

Fink's Class IV covers the most numerous examples of the true Roman lamp—those with plain nozzles having a line incised at the base, other plain nozzles, and those with heart-shaped nozzles. Mr. Walters (British Museum) not only subdivides the class into seven forms but places the heart-shaped nozzle in the third century as against the second-century date of most of the plain-nozzled types. A number of archaeologists disagree with this latter dating, and it seems to be decidedly discounted by the presence of heart-shaped nozzles in the finds from Pompeii (destroyed A.D. 78). Fink gives Class IV as dating from the first to the fifth century A.D., but most of the lamps in it are not later than the third century at any rate. Nearly all of them (especially the Italian specimens) have handles, but contemporary lamps from Greek sites are without handles and generally decorated with simple animal or conventional subjects. Second-century lamps from North Africa usually have simple designs, as altars, fruit, vases, &c.

The statement sometimes made that most Roman lamps were made in Italy and exported thence is a doubtful one. It seems likely, however, that the majority of the best glazed specimens are of Italian origin, though the repetition of certain motives in lamps from defined areas (as, for instance, a bird on a branch, occurring on several Roman lamps from Cyprus) suggests local manufacture. Attempts to determine the origins of Roman lamps by the fabric are not altogether conclusive. Birch, in his *Ancient Pottery*, refers to white lamps from the celebrated potteries of the Via Nomentana, dingy brown or yellow lamps from Naples, yellowish from Britain and Gaul, and pseudo-Samian lamps (there being no lamps in the true 'terra sigillata' ware) from Italy, but these dicta can hardly be accepted as conclusive except perhaps as to the last. Certain fabrics, however, seem to be peculiar to limited areas. For instance, the rather rare examples of lamps coated with green glaze seem to belong to the area round Naples, though one authority gives them as being made in the south of France. Black lamps and those of a pinkish-red ware are Romano-Egyptian. Generally speaking, extremes meet in that the coarsest wares are common to Britain and Egypt, especially in red-brown, unglazed red, and unglazed drab. Red and brown glazes are most common in lamps from Italy, though the red glaze, so far as the provenance of the lamps is concerned, is well distributed over the Empire. Roman period lamps from Cyprus are often of this fabric, as are also those from Ephesus. Britain shares with Gaul a yellowish-brown fabric seldom seen elsewhere; most unglazed lamps have come from the outer parts of the Empire. (Mr. Walters gives Greek sites as yielding lamps of a pale buff colour, Cyprus reddish-brown, Naukratis deep brick-red, Campania greenish-yellow or dull brown, France and Britain, if not imported, white or yellowish.) Broneer states that Greek lamps of the Roman era, his form XXVII, B.M. form 102, were largely made at Corinth.

An interesting group of Roman lamps is that in which the lamp itself, instead of being round with a design on the *discus*, takes the form of some animal or object. The most common of all is that of a human head (Seilenos or a negro usually), but other forms include a bull's head, a human foot (generally sandalled and with considerable detail of the footwear), and dolphins. These abnormal forms are particularly common among Romano-Egyptian lamps, and the heads and faces especially seem to have emanated from Alexandria. It has been suggested that they are imitations of bronze forms, but this is by no means proved and many of the clay examples may be earlier than the generality of bronze lamps of this type. Usually they have ring handles and are not later than (probably before) the third century A.D. in date.

Other 'abnormal' forms of lamps are multi-nozzled, up to fourteen 'burners' in some cases.

These lamps were often provided with bronze stands of varying heights up to about 5 ft. and some of them capable of being raised or lowered at will. Such standards take various forms, such as architectural features and trees with branches from which the lamps can be suspended (though bronze lamps were probably used for the latter in most if not all cases). Where they were used for ordinary hand lamps, the standards are provided with a flat platform on top.

One form of pottery lamp (form 70) was evidently intended to be hung from a nail or peg on the wall, as it has the handle at right angles to the axis of the lamp instead of in a line with it. Another form, which the British Museum classification does not dignify with a form number (a typical example being given as a 'variant of form 86'), had a stem in the centre which Broneer says was surmounted by a loop for hanging (see No. 2, Plate XII). The stem is almost invariably broken off, but Broneer illustrates broken-off loops obviously belonging to such lamps (Broneer, *Corinth*, vol. iv, part ii, Terracotta lamps). Most of them seem to be early (first century A.D.).

[*See Plates XII, XIII, and frontispiece.*]

ROMAN POTTERY LAMPS, BRITISH MUSEUM CLASSIFICATION
(Excluding abnormal forms)

Types with blunt or square-ended nozzles.
- (*a*) With ear handles at sides. Forms 72 to 74.
- (*b*) With handle in ordinary position or absent. Forms 75 to 77.

Lamps with pointed (splayed) nozzles with volutes. Early first century A.D.
1. Without handle. Forms 78 and 79 (the latter with 'ears').
2. With handle. Form 80.

Lamps with rounded nozzles with volutes. First century A.D.
1. Without handle. Forms 81 to 83.
2. With handle. Forms 84 and 85 (forms 86 and 87 with attachment to handle).
3. With handle and two nozzles. Forms 88 and 89 (with attachment to handle).

Lamps with depressed centre and long nozzle. First century A.D. North Italian fabric.
1. With three knobs on rim, no handle. Form 90.
2. With two knobs on rim but no handle. Forms 91 (with mask) and 94.
3. With handle. Forms 92 and 93.
   (Romano-British variants of above with interrupted groove.)
   See illustration for distinctive features of the above group (Fink's group III).

Lamps with plain nozzles and handles. Second century A.D.
1. With groove at base of nozzle. Form 95.
2. With plain nozzle. Form 96. Forms 97 to 99.

Lamps with heart-shaped nozzles. Third century A.D. according to Mr. Walters but undoubtedly going back to the end of the first century A.D. Forms 100 and 101.

Greek types of the Roman period.
1. With angular nozzle, handle, and Greek inscription. Form 102. (Usually with small panel on each side of rim.)
2. Plain or heart-shaped nozzle, no handle.
   (*a*) Nozzle small and plain. Form 103.
   (*b*) Nozzle with groove at base. Form 104.
   (*c*) Heart-shaped nozzle. Form 105.
   (*d*) Tarsus type. Form 106.
3. Late or quasi-Christian types with plain nozzle and solid handle. Forms 107 and 108.

A more or less normal late type has the handle at right angles to the axis of the lamp instead of in line. Form 70.

(*a*) ROMAN LAMPS IN ABNORMAL FORMS

(*b*) GREEK AND ROMAN BRONZE LAMPS

CHAPTER XI

CLASSICAL LAMPS IN METAL

The use of metal for the manufacture of lamps goes back to the latter portion of the fourth millennium B.C., the date assigned by Sir Leonard Woolley to the gold, silver, and copper lamps, apparently derived from the shell lamps in shape, found in Ur. Four copper lamps shaped like a kind of shell were included in a hoard of objects beneath a Sargonid Palace at Eshnunna and were dated at *c.* 2700 B.C. Even the late Roman idea of making metal lamps in animal forms was foreshadowed by a crocodile lamp of bronze from Ur. The Phoenician shell-form lamp was also represented by a large bronze lamp from Ras Shamra, *c.* 1000 B.C., and by bronze lamps found in the tombs of Cyprus—occasionally placed in niches or fastened to the wall by spike attachments. Other bronze lamps of this form found by the Wellcome Expedition in Palestine are dated at about the fifth century B.C.

Another and different example of an early metal lamp, or, rather, group of lamps, was the Hebrew seven-branched golden candlestick, the lamp character of which is shown by the ordinances for the provision of olive oil for refuelling set out in Exodus and Leviticus.

The bronzes from Luristan (the dating of which varies from 1600 to 400 B.C.) include lamps, one of which not only carries on the animal motives in the form of a head overhanging the lamp and a decorative design of ibex and lions on the bowl, but anticipates another late Roman feature by having a low stand or foot beneath the reservoir. A Luristan lamp (bronze) in the author's collection[1] is of a simpler type, having a square open bowl with four finger-like projecting spouts and a low foot at each corner of the bowl. A lamp illustrated in Layard's *Nineveh and Babylon* and given as 'Greek or Roman' is rather reminiscent of this latter Luristan type except that the reservoir is a round bowl; the long-channelled spout is like those on the Luristan lamp.

A few bronze lamps are extant which date from the Hellenic era—mostly more or less open—and the Erichtheum, on the Acropolis at Athens, was said to have had a golden lamp made by the sculptor Callimachus which was refilled annually (!) on the same day each year; obviously it was not in constant use. Bronze lamps are recorded as having been used in connexion with oracles at Pharae (Achaia) and Argos. There are bronze open lamps in the Alexandria Museum and a closed example with a spike for sticking in the wall.

An Etruscan lamp has a shallow bulged body and is set on a pedestal. Another, at Cortona, said to be Etruscan of the fifth century B.C., is a sixteen-nozzle lamp ornamented with sirens, satyrs, river gods, a mask of Medusa, and bands of animals in relief.

[1] No. 1, frontispiece.

It was in the Roman period, particularly from the first century A.D. onwards, however, that the use of metal lamps became more or less common, though always less common than that of the fictile forms, which were in general use in the poorer houses and for the commoner purposes of the household (as well as for funeral deposits—which accounts for many of them showing no signs of burning). The British Museum, in an attempt to group the forms employed in metal, catalogues a classification of nine general forms of Roman bronze lamps, but it is far more difficult to assign these to definite forms than is the case with the more standardized pottery lamps, owing to the wide range of variations and fanciful shapes the metal lamps—cast and finished on the lathe—evince. Many of the earlier examples show a distinct affinity with the regular pottery forms, especially the first century A.D. voluted-nozzle forms (the volutes on bronze examples being somewhat simplified); the divergent group in pottery lamps, forms 90 to 94, having the plain bowl and groove to the nozzle and particularly prevalent in Britain and Gaul is almost exactly reproduced in form 5 of the bronze lamps. Decoration on the bowls themselves, in the other forms, is rare and in its place there is generally attached extraneous ornamentation of some sort. In a fairly common group of lamps this takes the form of a 'loop' handle finishing in a horse's or other animal's head overhanging the body of the lamp, which is of an elongated pear shape (form 6). Another frequent though rather less common design has a figure of a god or goddess under a canopy or arch standing upright at the back of the lamp in place of a handle (B.M. form 1).

An important group (generally speaking later in date than those already mentioned) is that of bronze lamps made in various forms—human heads, animals, sandalled feet, dolphins, fir cones, even whelks and snails, &c.—which is sometimes said to have preceded and suggested the use of similar forms in the pottery lamps (more especially in Roman Egypt). Some of these are very elaborate, as the lamp in the form of a Seilenos in a chariot drawn by two bulls in the British Museum, and a bronze lamp, illustrated in the British Museum catalogue, in which the suspension chain is attached to two dolphins, heads downwards to the lamp, the latter being flanked by rampant lions. Even more elaborate are the fifth century A.D. baptismal lamp of Valerius Severus (a ship in full sail, with SS. Peter and Paul on board) in Florence Museum and the fourteen-light ship-like lamp found in Pompeii (now in Naples Museum).

Bronze lamps are frequently for many lights; in addition to the fourteen-light lamp already mentioned, another from Pompeii had nine burners. They were often provided with loops and chains for suspension and were more permanent domestic-lighting fixtures than the common earthenware lamps. A recent find in Herculaneum revealed a shop practically intact, with a bronze lamp hanging in place from the centre of the ceiling, no doubt the common position for such a lamp. Another quite normal position for a bronze lamp, however, was on a bronze candelabrum or lamp-stand, in which case it was either set on a flat disk at the top of the stand or hung by suspension chains from an arm (which sometimes took the form of the branch of a tree with the upright standard as the trunk).

A large proportion of the Roman bronze lamps which have survived the passage of time belong to the Byzantine or Early Christian era (from the fourth to the seventh century A.D.). These were usually for suspension or provided with an aperture at the base for the insertion of a spike on a stand; in both cases alternative standing use was allowed for by the provision of a circular foot. These lamps follow the general lines of the earlier lamps, though, as in the case of the pottery lamps, the Greek element reasserts itself. There is rather more elaboration in this period and the Christian connexion is frequently shown through the Cross, the Chi Rho, or other Christian emblems, by way of attachment, ornamentation, or engraving—the last-named usually on the base or on the hinged cover to the filling hole with which these lamps are generally provided. More rarely, the engraved emblem is on the side. Sometimes the lamps themselves take an emblematic form, a very common example of the epoch being that of the dove (see No. 3, Plate XIII, *b*).

Two very interesting discoveries of late bronze lamps of classical type were described in the *Illustrated London News* in June 1932, February 1933, and January 1936. Those described in 1932–3 were found at Balliana, south of Abu-Simbel, and were Roman-Nubian, of the fourth to sixth centuries A.D., a period when Nubia was being gradually christianized. The striking feature of the group is the mixture of paganism with Christian emblems. They included a standard in the form of Apollo, holding aloft two columns acting as lamp-stands, another with two lamps incorporating Maltese crosses held aloft by a pagan deity, a bronze lamp surmounted by a cross but supported by a figure of Eros (each of these examples being 2 ft. or more overall in height), and a fish or whale form lamp on an ornamental pedestal. The lamps described in 1936 were found at Firka, in the Sudan, and included, in addition to the common dove form, a lamp of fish form on a stand terminating in three leonine feet.

The classical bronze lamp does not seem to have survived or left recognizable descendants to the extent of the pottery lamp, but its influence may reasonably be traced in a Persian lamp of later date (see frontispiece) in the possession of the author as well as in a small early Arab lamp. The Persian lamp has the attachment at the back, the hinged cover to the filling hole, and the two projecting spouts of so many Roman or Byzantine examples, but the decoration is typically Persian, as is the bird on the filling-hole cover (again reminiscent, however, of the dove on the filling-hole cover of some Roman Christian specimens).

Occasional examples of iron lamps of the Roman period are found. These, though, usually have so much in common with lamps of barbarian origin as to suggest they are of hybrid parentage or indigenous to the country in which they are found; there is a distinct suggestion of the crusie lamp in some of those found in Britain.

A few examples also exist of leaden lamps, including one of the shell form from Cyprus (the author has a modern version of this type from Malta). A *golden* lamp from Pompeii is in the Naples Museum, but lamps in the precious metals must be regarded more as 'show' pieces than for actual use.

A word of warning to the would-be collector should be given here. Numbers of brass souvenir copies of Roman bronze lamps (especially Early Christian forms) and candelabra (the branching-tree example from Pompeii in particular) are sold in Italy and find their way into antique shops in England as 'Roman lamps'. They are easy to detect: the patina is thin and vivid, the designs stereotyped copies of well-known forms, and the modern brass is apparent when the patina is scratched. Unfortunately, at least nine out of ten 'Roman' metal lamps offered for sale are in this category and some of them have even found their way into local museums as genuine examples. (See Nos. 1 and 4, Plate XIII, *b*.)

Classical forms have appeared in metal lamps (especially in bronze) in the Renaissance period of Italy. One type (Florentine, seventeenth century) reproduces the Seilenos head, mounted on a stem terminating in eagle's claws at the base (see example in Birmingham Museum).

[*See Plate XIII and frontispiece.*]

CHAPTER XII

GRAECO-ROMAN LAMPS FROM EGYPT

No country has yielded so much material for the reconstruction of its ancient life as Egypt and, if the lamps of the native dynastic period are conspicuous by their absence (so far as they have been identified), this fact is balanced by the enormous number of pottery lamps of the Ptolemaic and Roman epochs found in that country. A large number of these are lamps of typical and normal Greek and Roman forms, but an even larger number are more or less peculiar to Egypt in form and design, though presenting a wide diversity of both in their details.

Leaving aside the unmistakably Greek lamps (forms 21 to 47, British Museum classification)[1] and the true Roman lamps, the essentially Egyptian types may be broadly divided into two large groups—those approximating more or less closely to Greek types (forms 48 to 55) and lamps of a more or less ovoid, convex-topped form of the 'frog-*motif*' type and types approximating thereto. Both groups are usually without handles, a feature which, in the case of the later examples in particular, marks a divergence of development from most of the contemporary lamps of the Mediterranean area which, while losing the neat little ring handle of the true Roman lamp, usually retain a stub handle or thumb piece at the back of the lamp or behind the filling hole; this occasionally, but not often, appears in the Egyptian lamps.

Dating the first group appears to be a matter of controversy. Sir Flinders Petrie, in *Roman Ehnasya*, seems to regard the lamps of this group as, in some cases at least, later than the 'frog' types, and as belonging generally to the third or fourth century A.D.[2] The British Museum catalogue,[1] on the other hand, puts some of them, especially the delphiniform types, as definitely of the first century B.C. Taken on the whole, the evidences in the lamps themselves, apart from any question of their provenance or the associations of their discovery, seem to be more strongly on the side of the earlier than the later dating. For instance, the nozzle forms of the group as a whole are distinctly Greek in style, being either lozenge-shaped (rarely), 'splayed' (a very typical late Greek form which is common in these lamps), or with upturned circular orifices.[3] Certainly the persistence of the Greek nozzle in Egypt later than elsewhere is not impossible, but it would be rather remarkable, in view of the fact that it disappears after the earlier part of the first century A.D. in other parts of the Roman Empire, and is not present in the ovoid frog and allied forms, in which indeed, in most cases, there is no markedly projecting nozzle at all. Again, all these Greek forms of the first group have long or longish 'necks' leading to the nozzle or

[1] Walters, *Catalogue of Greek and Roman Lamps in the British Museum.*

[2] *Roman Ehnasya*, pp. 8–9.

[3] These nozzle forms are typical of the second and first centuries B.C. in Greek lamps.

wick-hole itself, such as are more commonly associated with Greek than with Roman forms. On the other side of the case, there are one or two points which lend a certain amount of colour to the arguments for later dating. One is the fact that one form of lamp combines the frog *motif* with a projecting nozzle having a blunt terminal (though not so markedly Greek in form as the splayed and circular nozzles), and another is that, in rare cases (including one in the author's collection), such lamps have the stub or thumb handle which is usually a sign of a comparatively late lamp, not earlier than the third century A.D. (e.g. the lamp No. 8 on Plate I, 'The Bucheum', O. H. Myers). Taking everything into consideration, the author is inclined to the view that the majority of these lamps are Ptolemaic, though they may well have persisted, side by side with the true Roman lamp, well into the Christian era, just as the primitive saucer lamp appears in Romano-Egyptian excavations in the Bucheum (see No. 12, Plate X, *b*).

The second group may be more unreservedly ascribed to the third century A.D., with some later examples, as given by Petrie. This group is very distinctively Egyptian, with no parallels elsewhere, and whereas the lamps of the first group are usually red or black in colour, those of the second are most commonly white or grey—though the hybrid frog type with semi-Greek nozzle already referred to is more usually red.[1]

Already, before the third century of our era, Egyptian lamp factories seem to have delighted in abnormal forms, a large proportion of the 'face' lamps of Roman date being from them (Alexandria being suggested as a particular source), and it is possible that one of the commonest and most typical 'breakaways' of Roman Egypt was suggested, in the first instance, by the 'animal' lamps, inasmuch as some of these show an animal, not as the body of the lamp itself but as relief on the lamp (*vide* mouse example, No. 12, Plate XIII, *a*). Taking these as connecting links (and bearing in mind that the 'face' lamps also take the same course, as evidenced by Nos. 9, 11, and 6 on the same plate), there is a possible transition to the 'frog' type lamp, which Petrie attributes to the third century A.D. in the first instance. A possible objection to this reasoning is that Petrie dates the lamps showing the whole frog as later than the more numerous examples in which only the head and limbs are shown. While it is true that a number of lamps of the 'whole frog' type may be comparatively late in the development of such lamps, the natural order of things seems to suggest that some examples of this complete representation, at any rate, preceded the conventional abbreviation. (The 'frog' lamps with the Greek connexions are generally with part representations—sometimes the legs only—and these would appear to be early forms.) Incidentally, Professor Petrie remarks that it is strange that the frog never has the mouth forming the wick-hole; this, though, is in line with most of the 'abnormal' face and animal lamps

[1] Form J. Mr. Benachi suggests that the legs on these early lamps are those of monkeys, but on those in the author's collection the legs seem to be frogs' or toads', as in the case of later forms.

of the classical type, in which the spout or nozzle is almost invariably independent of the features.

Generally speaking, the frog lamps and their relations are more or less ovoid in form, but the Greek spouted form already mentioned is roughly 'D' shape, with a short spout projecting from the straighter 'front' of the lamp. They are usually decorated with what Petrie describes as the 'corn' *motif* and constitute his 'frog and corn' group. The frog itself, it is suggested, was used as an emblem of resurrection; might it not have been emblematic of the great river which was and is the very life-blood of Egypt? Some of the lamps, indeed, seem to have the water symbol incorporated.

In *Roman Ehnasya* Petrie has drawn up a detailed classification of Romano-Egyptian lamps. Such classification is not an easy matter with the numbers of differing but apparently contemporary types among the thousands of such lamps in existence (and the numerous 'crosses'); they are probably the commonest of all pottery lamps extant, not even excepting comparatively modern survivals of primitive forms. Petrie enumerates, in all, 25 groups (based on over a thousand examples) plus a number unclassed; excluding lamps which are, or approximate to, true Roman lamps or which show extraneous influences, the markedly Egyptian groups of his classification number sixteen. Even these, however, include a number of lamps of 'dolphin' (delphiniform) type which may belong to the Greek classification of Mr. Walters, but which Professor Petrie (as already stated) connects with types which he dates as third or fourth century A.D. With all due respect to so eminent an authority, the Petrie system of classification seems far from satisfactory, based as it is largely on decorative motives, with the actual basic form of the lamp relegated, in many of the groups, to a secondary or even non-essential position and the form of the nozzle (generally a significant feature in detecting origins and dates) totally ignored. Such a system means that lamps of similar form are put into different groups and that some of the groups contain lamps of widely differing forms divorced from their relations. It also runs completely counter to the methods adopted by Mr. Walters and by Herr Fink for Roman lamps and thus puts out of court any co-relation of the classifications.

The author's first section is more or less covered by Petrie (who dates most of them to the fourth century A.D.) under his categories 'W' (Wreath)—some of his examples in this group being of true Roman form—'K' (Echinus), some of which, according to his illustrations, have the ovoid body of the frog type, while others, the majority, are Greek in form—'O', Round-bodied, 'S', Shouldered, 'V' (Delphiniform), 'U', Radiate, and 'L' (Loop), groups 'S', 'V', and 'U' being fairly homogeneous. Most of the lamps have long nozzles and are in red ware or (less frequently) black. They have round or 'D' shaped bodies and convex (domed) tops of late Greek type.

The author's second section covers lamps never found outside Egypt. It includes the frog types and a number of other lamps of similar general form (ovoid, with convex top), usually in a grey or drab clay. Professor Petrie classifies most of them under 'F' (Frog lamps), 'E' (frog and corn), 'D' (corn and palm), 'J' (joint types), 'A' (arm), 'B' (boss), 'Y' (deep-cut), and 'X' (Ankh and Cross). The last-named usually have an incised cross of the 'Tau' type—a Coptic form which seems to be derived from a confusion or amalgamation of the cross and the old Egyptian 'ankh' symbol of everlasting life.

A third and late section covers a number of elongated pear or shoe-shaped lamps with a groove to the nozzle which are closely analogous to Syrian and Palestinian types, but are usually rather narrower in proportion to their length than the latter. These are commonly in brown or red ware, unglazed. They are both later and less typically Egyptian than the first two sections.

With all diffidence, and bearing in mind the objections to the Petrie *motif* method of classification, the author would venture to put forward an alternative general classification for the typically Egyptian lamps of the Graeco-Roman era, viz.:

A. Flattish round body, long nozzle with splayed terminal. (Generally, Sir Flinders Petrie's round-bodied types.) No. 1, Plate XIV.

B. Double convex body, long nozzle, splayed. (Usually of the Petrie 'radiate' type lamp.) A common form of the Greek section. No. 2, Plate XIV.

C. 'Delphiniform' lamp with single projection at side, double convex body, splayed nozzle. (This can be related to Walters' form 48.) Commonly in black glaze. No. 3, Plate XIV.

D. Flattish body, projection at each side, generally with splayed nozzle, often with 'wreath' *motif*. (Petrie's 'spur' form.) Sometimes in black. No. 4, Plate XIV.

E. Flattish round body with 'shoulders' at end next to nozzle; nozzle splayed or plain. (Petrie's 'shouldered' form.) No. 5, Plate XIV.

G. Pear-shaped double convex body, usually (but not invariably) with rounded nozzle. 'Echinus' *motif*, with palmette behind nozzle. No. 6, Plate XIV.

H. Shallow 'D'-shaped body, slightly convex top, round nozzle, with volutes. No. 7, Plate XIV. Probably Roman, First century A.D.

The forms following are of the second section, including the frog types:

J. Double convex body (round) with a shorter nozzle (blunt ended) than the more purely Greek forms. This form is usually, but not invariably, in red or brown ware. No. 8, Plate XIV. (The shape of the nozzle suggests an earlier date than the rest of this section.)

K. Flat 'flask'-shaped body, very little projection to spout, 'corn and palm', 'arm' and 'frog' *motifs*. No. 9, Plate XIV.

L. Rounder and deeper body than 'K', convex top, slightly projecting spout, 'frog' *motif*. No. 10, Plate XIV.

M. Similar shape, 'corn and palm' *motif*, with knobs, on convex top. No. 11, Plate XIV.

N. Similar shape, with knob ornament only. No. 12, Plate XIV.

O. Pear-shaped, no projection to nozzle, which is a simple hole at the narrow end. Sometimes with 'frog' *motif*. No. 13, Plate XIV.

P. Ovoid, no projection to nozzle, 'frog', 'corn and palm' motifs. Very numerous. No. 14, Plate XIV.

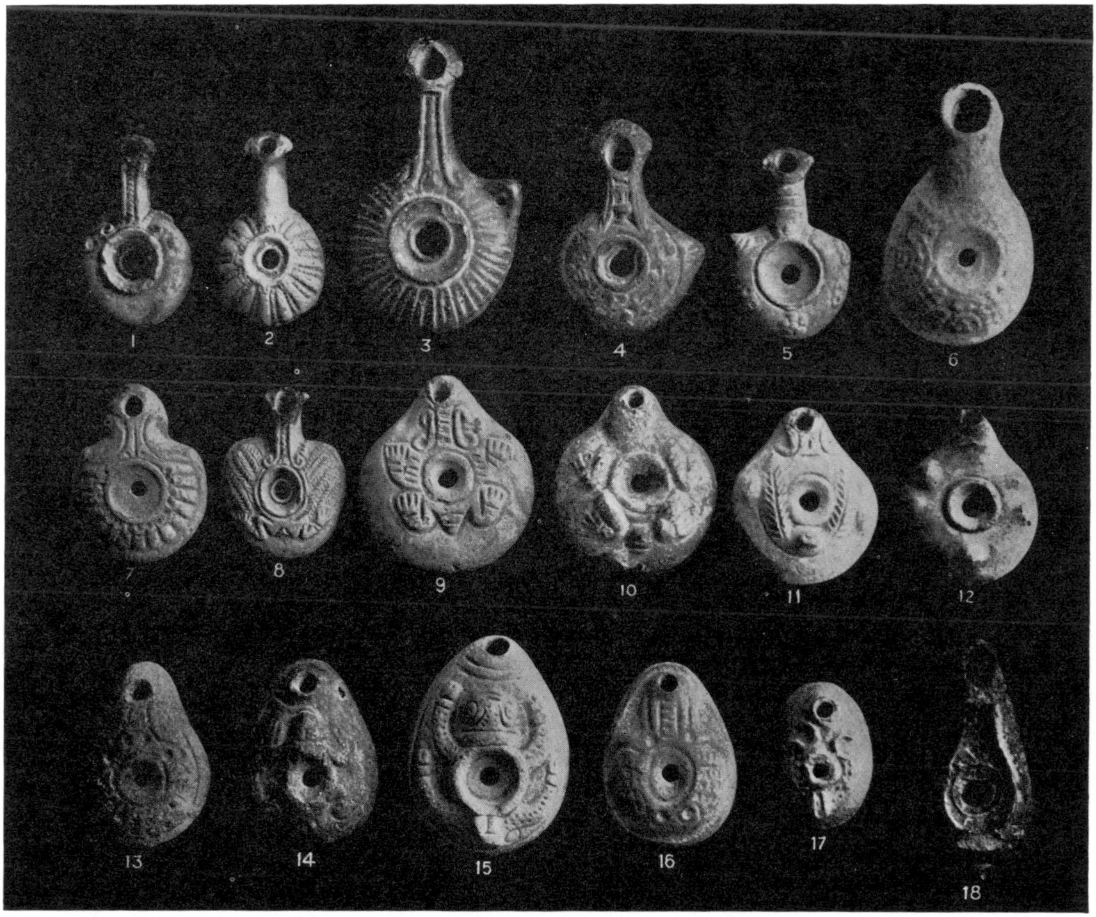

GREEK AND ROMAN EGYPTIAN LAMPS

Q. Ovoid, with raised rim around filling hole and stub handle behind latter. Petrie's 'Ankh and Cross' type, but also includes deep-cut examples with frogs. No. 15, Plate XIV.

R. Narrow ovoid body, no projection to spout. Decoration includes degenerate frogs, faces, and other *motifs*. No. 16, Plate XIV.

S. Quite oval form of body, double convex, no projection to spout, no decoration except for 'C'-shaped ridges near nozzle, which may be crude attempts to copy volutes. No. 17, Plate XIV.

The following form constitutes the third section:

T. Usually almond or shoe form, handled, groove to nozzle; a late type running up to the seventh or eighth century A.D., and connecting with Syrian types, &c. No. 18, Plate XIV.

All these lamps are moulded. For illustrations see Plate XIV.

A group of lamps which has not been taken into consideration and of which specimens are rare is one in which the form is that of a teapot or pitcher, with a flattish front perforated with holes, and a spout projecting from the lower part. This group Dr. Grenfell considered Ptolemaic and Mr. L. A. Benachi (Alexandria)—who has a notable collection of Egyptian lamps—informs me that not only do the clay and finish of these lamps point to a Ptolemaic date but they have actually been found in some Ptolemaic tombs. Also, they are wheel-made (another suggestion of early date) and, unlike the lamps dealt with above, are handled. The high body, rimmed top, and handle, however, are points of resemblance to seventh to eighth century lamps of beehive type on the one hand and to Assyrian and Parthian lamps on the other (see Plate XV, *b*, for all these), so that the author is inclined to suspect a West Asiatic relationship.

CHAPTER XIII

'LATE' POTTERY LAMPS

THE pottery lamp of classic type was at its best, both as to design and fabric, during the first two centuries of the Christian era. After that, there set in one of those declines which are illustrative of the fallacy of considering progress as something continuous and unchecked. The decline is not particularly marked in the lamps of the third century A.D., but the first signs of carelessness creep in. An outstanding and dating feature is the appearance, in place of the neatly made ring handle of the earlier Roman lamps, of the solid handle, fairly well-made of its sort in the first instance, but gradually deteriorating into a mere thumb projection, on the one hand, or, on the other, to a carelessly made affair of varying degrees of crudeness and often, in its late forms, not even in a true line with the axis of the lamp. This solid handle is a sign of late work—from the third century A.D. onwards—but particularly during and after the fourth century.

Many of the lamps of the third century from sites in the Roman Empire are, so far as the body of the lamp is concerned and apart from the solid handle, true Roman in type, having the round body, moulded rim, depressed bowl with more or less classical design and small filling hole. Reference has, however, already been made to the peculiarly Egyptian types of that period and a similar process of divergence from Roman forms occurs in other eastern Mediterranean areas.

Despite the superior design of the Roman lamp and the fact that it percolated into all parts of the Empire, Greek influence appears to have persisted in the Near East and lamps found in Syria, Palestine, &c., dating from the third century A.D., are either hybrids between the Roman and Greek lamps or markedly Greek in type, though the mould method of manufacture, as distinct from the Greek wheel-turning, rendered unnecessary the adherence to the round form. The large filling hole and convex top (in place of the depressed bowl with small filling hole) eventually prevailed altogether. Lamps of this type in the third and fourth centuries are more or less circular in form, with conventional floral, cable, &c. ornamentation around the rim or behind the nozzle; later they are elliptical, almond, pear, or shoe-shaped.

Excavations in Beit Nattif, Palestine, in cisterns built in the first century but used for pottery deposits during the third, yielded three types of lamps—large buff ware lamps with flat tops and small filling holes, having Roman designs, small red painted lamps with rounded tops having larger filling holes and simple conventional designs, and rather similar lamps having Jewish or Christian symbols at the nozzle end. No doubt these are all close in period—possibly contemporary—but they illustrate, in order, the trend of the later lamp.

The next stage in the recession from the true Roman form is the gradual departure from the rounded form and slightly projecting nozzle towards an ovoid and afterwards pear- or shoe-shaped form. This is the form common in Syria, Palestine, and Asia Minor (with examples in Egypt) from the fifth to the eighth centuries. These lamps are almost invariably provided with solid or stub handles and are of far inferior workmanship and pottery to those of the lamps of the first two or three centuries of the Christian era. In colour, they range from white or grey to buff and pink and are decorated generally with geometrical or conventionalized floral designs, sometimes with Christian emblems. A design fairly frequently found in Palestine shows a symbolical representation of the seven-branched candlestick, probably as an emblem of race or faith.

In these later lamps the formed nozzle has usually degenerated to a mere hole in the narrower end of the lamp and, in the latest lamps, a more or less well-marked groove extends from the large filling hole to the wick-hole, which groove was probably intended to cause any little oil spilt around the filling hole to gravitate towards the wick—a provision which was clearly unnecessary in the better-designed Roman lamp, in which any oil outside the filling hole would naturally gravitate into the reservoir itself.

These fifth- to eighth-century lamps are, with the Romano-Egyptian types, amongst the commonest survivals of early pottery lamps and may be readily picked up, for a few shillings apiece, in curio shops in this country.

The diversity of types among more or less contemporary lamps and the uncertainty of dating makes it rather difficult to lay down a definite and comprehensive classification of these immediately post-classical lamps, but here, at any rate, is an attempt to do so:

I to IV. Beit Nattif types.

I. Round, pseudo-Roman in type, flat or depressed bowl (discus), small filling hole, small stub handle, formed round nozzle with cable design behind. Large buff examples at Beit Nattif.

II. Round, depressed bowls with small filling holes, geometrical ornamentation and borders, broad nozzles. Large unpainted examples at Beit Nattif.

III. Round, semi-open (large filling hole) double convex bodies, broad nozzles with interlaced work behind, stub handles. Small red-painted examples at Beit Nattif.

IV. Round, semi-open (large filling hole) double convex bodies, small stub handle, slightly projecting nozzle with cable design behind. Small red-painted examples at Beit Nattif. No. 2, Plate XV, *a*.

V. (A similar type.) Round body, double convex, semi-open (large filling hole). Slightly projecting spout, plain nozzle.  behind nozzle. No. 1 on Plate XV, *a*.

VIII. Slightly oblong, nearly flat, rimmed filling hole, solid 'D'-shaped handle, groove to plain nozzle. Type 'A' British Museum Early Christian catalogue. No. 5 on Plate XV, *a*.

IX. Round lamp, open centre to bowl, pierced solid handle. Type 'B', British Museum catalogue.

X. Nearly oblong in shape, closed lamp, small filling hole, rather splayed nozzle, solid handle. Type 'C', British Museum catalogue of Early Christian antiquities.

XI. Deep nearly round body, upright solid handle, closed in top, groove to wick-hole. Type 'E', British Museum.

XIII. Almond-shaped closed-in body, depressed bowl with convex top, large stub handle, small filling hole, plain nozzle.

XIV. Similar to above but ovoid in form, large filling hole and almost horizontal solid handle. No. 11 on Plate XV, *a*.

XV. Nearly round body, projecting blunt nozzle, stub handle behind body of lamp.

XVI. Almond or shoe-shape, no handle, large filling hole, no channel or groove, conventional representation of seven-branched candlestick on front of lamp. No. 10 on Plate XV, *a*.

XVII. Ovoid, double convex body, small depression around filling hole, stub handle behind latter (not at back of lamp), top decorated with foliate ornamentation. No channel or groove. No. 12 on Plate XV, *a*.

XVIII. Ovoid, small filling hole, small stub handle behind filling hole (not at back of lamp), shallow groove or channel to nozzle hole (which is not at the extreme end of the lamp).

XIX. Almond-shaped double convex deep body, small stub handle behind small filling hole, depression around latter, decorated sides, channel or groove to nozzle. No. 13 on Plate XV, *a*.

XX. Double convex body, 'D'-shaped solid handle, decorated sides, large filling hole, stub handle at back of lamp, channel or groove to nozzle. No. 14 on Plate XV, *a*.

XXI. Oblong lamp, large filling hole, double convex body, crude geometrical ornamentation, stub handle at back of lamp.

A rather common, globular form, with 'stippled' ornamentation and stub handle, is not included in this classification as it does not appear to belong to the area to which the foregoing lamps generally belong and is considerably more widespread. (No. 4, Plate XV, *a*.)

Another large group, as belonging to an area farther west (North Africa, especially the Carthage district, South Italy, and with examples as far north as Britain), is also omitted from this classification. These last-named lamps are an early Christian group, with oval bodies, deeper in proportion than the true Roman lamps, having flat or slightly depressed tops, upright solid handles and projecting grooved nozzles. Generally they are in red or brown ware, occasionally buff, and often unglazed. On account of the large number found in that area and its central position in the area of their distribution they may be conveniently labelled 'Carthage' type lamps, though their actual provenance seems to extend from Asia Minor and Egypt in the east to Britain in the west. Possibly Carthage was the centre of their manufacture, especially in the characteristic red ware found so prolifically there, but there is a very large range of the lamps of this type in Alexandria Museum. They are closely

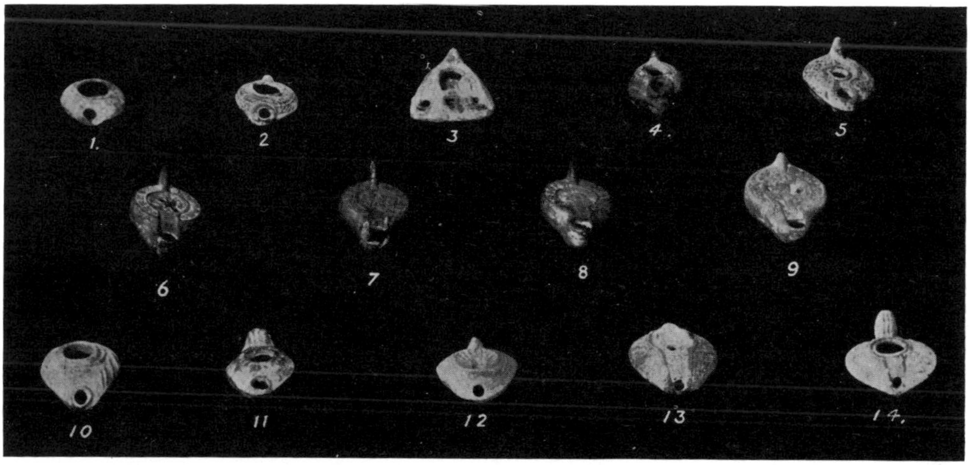

(*a*) LATE ROMAN, ETC., POTTERY LAMPS

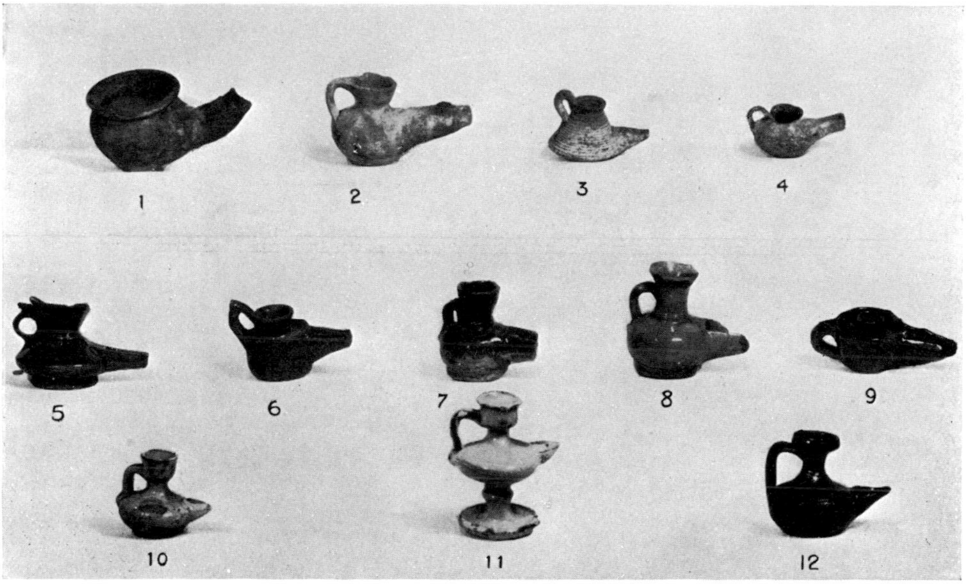

(*b*) ARAB AND RELATED POTTERY LAMPS

associated with the Christian faith, the great majority of the designs on the bowls being drawn from Christian symbolism. (The Alexandria examples show, in addition to the cross, palms, &c., the cock, the lion, and even Imperial busts, but, as a few of these lamps have ring handles they probably cover an earlier period than the general run of the type.) Usually, the main design is enclosed in a border of geometrical designs, palm branches, &c. (in one case at least, heads of apostles); as to the main design itself, the fish, the anchor (for hope), the good shepherd, the dove (the soul or the Holy Spirit), and the palm branch (victory) are among the earlier symbols drawn upon, followed by the peacock (immortality), the phoenix (resurrection), the palm tree (paradise), the vine, the triangle (the Trinity), figures of saints, martyrs, and biblical characters, and above all the Chi Rho or Christ monogram, which, used frequently from the fourth century onwards and adapted to form the monogrammatic cross (common after the middle of the century), preceded the cross itself in general use and survived side by side with the now universal Christian emblem until at least the seventh century of our era. Human figures represented on these lamps are frequently given the nimbus or halo which originally implied a person of importance (not necessarily a saint).

These lamps (middle row, Plate XV, *a*) range from the fourth to the seventh century A.D.

To Byzantine times (from about the fourth century A.D. onwards) belongs a lamp of the eastern Mediterranean which has the shape of a shallow bowl with a cone or beehive-like receptacle inside it, to which the sides of the bowl form a rim; the inner receptacle has a small hole in the side as a nozzle and a filling hole at the top which is surmounted by a funnel in the developed specimens.[1] About the eighth century A.D. (possibly earlier) another form appears in the same area, like a beehive in shape, with a nozzle near the base and the upper part tapering into a truncated cone which forms a funnel for filling (No. 3, Plate XV, *b*). This lamp seems to have been derived from Assyrian and Parthian lamps (Nos. 1 and 2 on the same plate) having globular bodies, rimmed filling holes at the top, and nozzles projecting from the base, with possible cross influence from the Byzantine lamps mentioned above, unless indeed these latter have also been affected by influence from an eastern source. The eighth-century lamp is a half-way step between the Assyrian-Parthian lamps and the early medieval Arab lamps (generally in blue or green glaze), varying in form from a circular body with flat top and bottom to a flattish spheroid section, and proclaims itself as an Arab importation and the prototype of the medieval lamps. All of the last-named are provided with a splayed funnel on top for filling, a long, projecting channelled spout, and an upright handle.

[1] A Persian medieval green-glazed lamp in the New Barnet Folk Museum has the beehive set in a shell-form surround but the nozzle-hole has become a mere depression and the funnel seems to be the wick-hole.

They seem to range from about the ninth or tenth century A.D. to the twelfth in these particular forms and ware, but that they have had a continuous lineage through the succeeding centuries until modern times is shown by the existence of lamps very similar to them in form in Morocco and elsewhere in the Nearest East to-day. Somewhat similar lamps are recorded by the Smithsonian Institution as being used in Turkestan, both in medieval and modern times, and they probably followed the spread of the Moslem faith, though particularly prevalent in Egypt. The spread of lamp forms is everywhere connected with the spread of racial or religious cultures, and marked ethnological evidences add interest to the study of the story of this intimate domestic utensil.

Strikingly similar to some of the medieval Arab lamps in form are the white (or sometimes brown) glazed lamps of Apulia, South Italy—a similarity which suggests a descent from Arab lamps which is confirmed by the history of the peninsula, occupied in part by the Saracens at the very period when the Arab medieval pottery lamp was in vogue (the tenth century). These Apulian lamps are globular, with funnel tops, handles at the back, and small projecting leaf-shaped spouts. The type, like the brown glazed hand lamps of Morocco referred to above, is usually on a foot or low pedestal but is also sometimes on a tall stand or pedestal like the Neapolitan, Moroccan, and other pedestal lamps of the Mediterranean area. An unusually large hand lamp of this type in Ilfracombe Museum has a large shield (in pottery) behind the nozzle as though to shield the hand from the flame. (See Nos. 10 to 12, Plate XV, *b*.)

A somewhat similar lamp to Nos. 10 and 12, with the funnel top, seems to have been in use in Dalmatia, the author having obtained three specimens, in brown and black pottery, from Dubrovnik (Ragusa), but it is difficult to say whether these were derived from Italian contact or direct from Saracenic influence.

A type of lamp which does not seem to have any direct connexion with Eastern forms but which may be conveniently dealt with here is a globular-bodied lamp on a low pedestal from modern Greece. This has a top filling hole and projecting spout and is made in red clay with a black glaze, obviously copied from the red and black ware of ancient Greece (see frontispiece). Another Greek peasant lamp of somewhat similar form in the possession of the author is in brown glaze.

[*See Plate XV and frontispiece.*]

CHAPTER XIV

FLOAT-WICK LAMPS

THE beginnings of the float-wick lamp are lost in obscurity. It is likely that it was in existence long before the first known glass lamps of the 'sanctuary' genus appeared in the fourth century A.D. and that its birthplace was Egypt. Herodotus refers to saucer-like lamps in which the oil floated in water (a regular feature of the float lamp) being used in a festival of lamps at Sais and elsewhere, and the lamps shown in the wall paintings in the Rock Tombs of El Amarna are distinctly floating-wick lamps, whereas the Phoenician and early Greek types are provided with lips for the wicks. Circumstantial evidence is provided by the occurrence of the float lamp in the Fayoum in the fourth century, thence spreading in Byzantine times (in which so much that was Egyptian was adopted in the Eastern Empire and even crept into western Europe), and its adoption by the early Christian Church, whose ceremonies and furnishings are strongly tinged with the influence of Egypt—naturally so, since so much of the Christian ecclesiastical rule originated there (including the monastic system, destined to play such an important part in the subsequent development of the Church). In fact, it was probably the association of the floating-wick lamp with the early Christian Church and the internationalism of the monastic orders, more than any other suggested cause, that spread the lamp of the 'sanctuary' type over Europe, just as the Mohammedan conquest of Egypt was probably responsible for its adoption in the mosques of the Near East.

In the *Journal of Egyptian Archaeology* for November 1931 there is an exhaustive account of Byzantine glass lamps by Grace M. Crowfoot and D. B. Harden. By comparison with the descriptions of standing and hanging lamps of glass given by ancient writers, from the fourth century onwards, and the lamps depicted in mosaics of the fourth century A.D., the authors (and others quoted by them) have been able to identify a number of surviving vessels or fragments of vessels in glass, of Byzantine date, as lamps. Judging by types found at Karanis in the Fayoum and elsewhere, the earliest recognized forms (fourth and fifth centuries A.D.) were simple cones (either with a pointed or a narrow flattened base) or hemispherical. The analogy of the bowls containing flames seen in ancient Egyptian tomb paintings seems to suggest that the latter were the older type, and the difficulty of identifying so common a form of vessel with a particular use may explain the obscurity in which early Egyptian lamps are buried. Excavations among Byzantine churches of the fifth to eighth centuries A.D., at Jerash, Syria, yielded fragments of several different types of glass lamps, including beaker-shaped vessels with flaring mouths, bowl-shaped vessels on either solid or hollow stems, and three different shapes of three-handled vessels. Basing on these and other examples, the writers of the article proceed to classify early glass lamps of the floating-wick type into four groups: (1) conical glass lamps of the Karanis type (which they date as fourth to sixth century A.D.);[1] (2) bowl lamps, first plain and later with a stem below the base, the former similar to an example in the Treasury of St. Mark, Venice (probably seventh to eighth century A.D.), the latter found on several sites of the fifth to eighth centuries, persisting to the thirteenth century A.D. at least in Europe and used in the churches and mosques of Palestine to the present day; (3) handled bowls such as were found at Jerash (fifth to eighth century A.D.) and from which it is suggested developed the mosque lamp with tall, outsplayed neck, bulbous body, and base ring or high foot; and (4) handled cups with stems.

One cannot but be struck by the resemblance of some forms of Group 1, thistle-like, with splayed mouth and a knob at the base (occurring as late as the Middle Ages) to certain 'cups' recorded by Layard as having been found at Nimrod (illustrated on p. 600, *Nineveh and Babylon*).

The float-wick lamp, in its various forms, started its definite historic career as a single light (though Egyptian prototypes sometimes show more than one flame), and that it was not originally of exclusively ecclesiastical purpose is shown by the discovery of simple glass cone-shaped lamps for floating wicks, dating from the fourth century A.D., in dwelling-houses at Kôm Aushim (Fayoum). Some were plain and some decorated. As early as the sixth century A.D., however, it had been adapted as a component in a chandelier device, an idea which was adapted and incorporated in the elaborate chandeliers (misleadingly described as lanterns) in the Arab Museum at Cairo. Early float-wick 'chandeliers' had the lamps inserted in holes in a suspended disk.

The float-wick lamp, especially in its glass form, spread from its probable birthplace, Egypt, and its adopted home, Byzantium, westwards and eastwards. Its spread westwards was unmistakably fostered by its ecclesiastical associations, and no doubt the monastic system, originating as it did in the Egyptian desert, had a good deal to do with it.

Anglo-Saxon glass bowls with 'flared' or splayed tops found at Faversham and now in the British Museum seem certainly to be float-wick lamps; a splayed-mouth pottery bowl of the seventh century A.D. found at Bourton-on-the-Water and also in the British Museum is likewise strongly suggestive of a float-wick lamp.

The type survived into the Middle Ages and is depicted in several illuminated manuscripts of the twelfth and thirteenth centuries. While these documents are of ecclesiastical origin, the pictures into which the lamp is introduced suggest that its use was not confined to churches. MS. Bodl. 602 (British Museum) shows it lighting the work of a monk in the scriptorium and it has to be remembered that, just as the tallow dip was good enough for the conventual buildings while the more expensive wax candle served the altars, so the lighting of the ordinary monastic chambers would be more or less that of the better-class

[1] Cf. modern example, No. 5, Plate XVI, *a*.

(a)

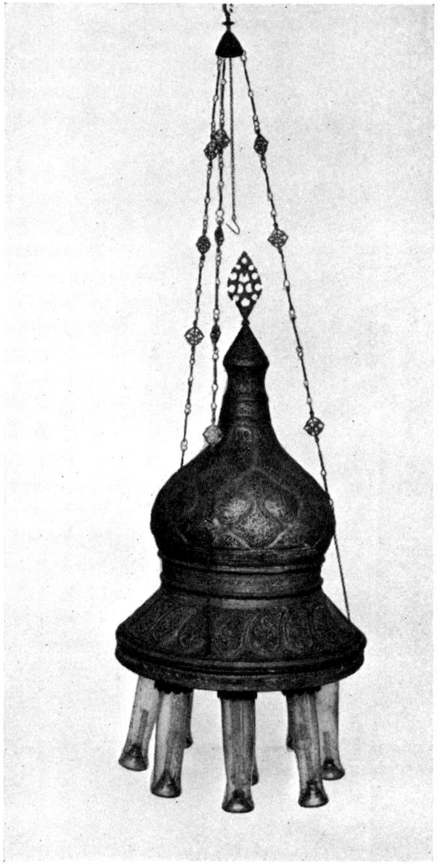

(By the courtesy of the Science Museum)

(b)

FLOAT-WICK LAMPS

homes of the period. Similarly, the appearance of the float-wick lamp in manuscripts depicting the Virgin and Child has to be taken in conjunction with the fact that it is an ordinary medieval domestic scene that is being pictured in the medieval illustrations, with the ordinary dress and habits of the period. Curiously enough, these appearances of the glass floating-wick lamp cease in subsequent centuries. Why? It undoubtedly continued to be used in churches (having survived for such use until to-day), and the fact that, notwithstanding this, it is absent from later medieval pictures in the manuscripts emphasizes the more general use in the earlier period. In view of the widespread use of the float-wick lamp in the East, one would have thought that the opening up of eastern trade through Venice would have brought the type, in more varied forms, still more generally into use, just as the eastern socket candlestick came along that channel, but the reverse seems to have been the case in western Europe at any rate, and while it has survived even for secular use, apparently, in some southern areas (the author having been informed, for instance, that float-wick lamps were used in early railway carriages in Spain), its principal place of use to-day is the church sanctuary, where it survives by force of tradition and symbolism. Sometimes, in that quarter, it is in its early form as a glass lamp, unobscured, but often now it is a glass reservoir in a more or less elaborate metal casing or holder. These metal pendant holders (frequently out of all proportion to the size of the lamp itself) were apparently in use, in gold, silver, or baser metals, in the Middle Ages and even as early as the ninth century of our era but attain their height of luxuriance in the Renaissance period. Notable examples are the silver lamp-holders of St. John's Cathedral, Valetta, Malta.

Isolated examples of glass standing lamps for a floating wick exist, as an example described as British in the Science Museum, South Kensington, and one in the possession of the author; the latter has a transverse 'bar' to support the wick, the ends of the bar fitting into notches on either side of the lamp. Most of the glass standing lamps (eighteenth and early nineteenth century), however, were fitted with a 'stopper' holding a central wick and were not therefore float-wick lamps.

A break-away from the usual medium (glass) is shown in a pottery lamp of the 'stemmed' base shape in the Ashmolean Museum at Oxford (early medieval in date and therefore contemporary with the glass forms in western Europe). Other early pottery lamps (some of Anglo-Saxon date) seem to have been intended for float wicks (see the chapter on medieval lamps).

In the East, probably through the double contact with Egypt and Byzantium, the glass float-wick lamps became accepted fittings, especially in palaces, mosques, and synagogues, and have continued in widespread use to the present day, though in North Africa they are now often to be seen minus their oil and fitted for electric-light bulbs! It was in the East that their decoration and embellishment developed most—on the one hand by colouring, incrustation, or enamelling of the glass itself, and on the other by the provision of elaborate metal containers or holders (sometimes the actual lamp was inside a glass bowl, just as the alabaster 'lamps' of Tut-ankh-amen's tomb probably contained smaller float-wick bowls). The enamelled glass lamp is a special feature of Arab lamps of the medieval era and several very fine and elaborate fourteenth-century specimens (resembling in form a thistle bloom on a splayed base and provided with loops for hanging) are to be seen in the British Museum.[1] The best collection is in the Arab Museum at Cairo and ranges from the thirteenth to the fifteenth centuries. These are the lamps referred to above as having developed from the handled bowls of Jerash, but they are lamp shades rather than actual lamps, a small saucer lamp vessel being placed inside, suspended by wires from the rim of the glass vessel. Medieval examples are decorated with enamel and inscribed with dedicatory or pious expressions.

In the Near East the bowl form is common and is often of pottery or even of metal. The glass lamps, in addition to the older forms and especially vase-like forms, developed a cylindrical form with a punched-up base; this has a slender glass tube inside to which a wick is attached (the wick in Palestine, for example, being a wisp of cotton wound round a strip of reed standing in the glass tube) and are usually suspended directly from an ornamental dome. It is interesting to note that the form of the lamp reservoir itself was foreshadowed by a similar form of Byzantine date. Similarly, the bowl, bell, or cone-shaped lamp with splayed top for holding and a knobbed base is Byzantine in origin. This is a common form in North Africa, both as a single hanging lamp and in chandelier constructions. It has obviously influenced the glassmakers of the West in comparatively recent times; vases on old-fashioned 'epergnes' could often pass quite well for glass lamps of the Byzantine type.

In Palestine the float-wick lamp of the Jewish home is a simple bowl with a float, often for several wicks—one in the centre and usually six around—and is lit every Friday.

The stone cresset lamps of early medieval date are, in principle, standing float-wick lamps. They nearly all show a small depression in the bottom of the bowl for the insertion of an upright to support the wick. Such a device (already mentioned as belonging to the cylindrical lamps of the Mohammedan world) appeared at an early date in the glass lamps and is common in Palestinian examples, alternating with a strip of metal bent into 'S' shape, one end being hung over the rim of the glass and the other folded upwards in the centre inside and bent at the end to hold the wick, thus kept down in the oil. An alternative in more modern float-wick lamps is a strip of metal traversing the width of the lamp, with a holder for the wick in the centre (as in the standing lamp of glass in the author's collection, mentioned above). Some such device may have appeared in some of the stone cressets as there are grooves suggesting it in a

[1] See modern *plain* blue glass specimen, No. 7, Plate XVI, *a*.

possible stone cresset in Wells Museum. Generally speaking, however, the float-wick lamps of to-day, other than the near eastern cylindrical variety, are provided with wicks carried in a floating disk of cork and tin; these wicks are sold in boxes and are sometimes used on the continent of Europe as 'night lights' for secular purposes as well as in the sanctuary lamps. Reference has already been made, in the chapter on saucer lamps, to the 'lamparistas' or float-wick saucer lamps used in Brazilian kitchens.

Though in appearance and use it is rather more related to the 'pan' lamp of Europe, an Ashanti iron lamp in the form of a single shallow bowl provided with a loop and a spike for sticking into the mud wall of the hut, made by a native smith but said to owe its being to Mohammedan influence from the north, is probably a decadent relative of the Moorish float-wick lamp. It is used with shea butter and a wick resting on the edge (No. 12, Plate XXIV, *b*).

A more truly float-wick lamp apparently occurs in New Guinea, in the form of a coco-nut bowl with a floating wick (*vide* 'The Dancing Dead', by R. M. Macdonald, *Cornhill Magazine*, December 1935).

In conclusion, not only did Victorian 'epergne' glasses suggest the influence of the form of the Byzantine glass lamp but many of our modern gas and electric-light fittings do also. This was notably the case with the splayed-mouth globes for the old-fashioned flat flame and upright incandescent gas-burners, but inverting some of the present-day shades will show almost equally Byzantine forms.

[*See Plate XVI.*]

CHAPTER XV

THE HANUKKAH LAMP

THE seven-branched candlestick of the Hebrews was undoubtedly, in its original form, a group of 'float-wick' lamps. This is obvious from the injunctions in the Old Testament. Exodus, chapter 25, verse 31 on, says:

'And thou shalt make a candlestick of pure gold: of beaten work shall the candlestick be made; his shaft and his branches, his bowls, his knops and his flowers shall be of the same. And six branches shall come out of the sides of it; three branches of the candlestick out of the one side, and three branches of the candlestick out of the other side; three bowls made like unto almonds, with a knop and a flower in one branch; and three bowls made like almonds in the other branch, with a knop and a flower; so in the six branches that come out of the candlestick. And in the candlestick shall be four bowls made like unto almonds, with their knops and their flowers. And there shall be a knop under two branches of the same, and a knop under two branches of the same and a knop under two branches of the same, according to the six branches that proceed out of the candlestick. Their knops and their branches shall be of the same; all of it shall be one beaten work of pure gold. And thou shalt make the seven *lamps* thereof; and they shall light the lamps thereof, that they may give light over against it. And the tongs thereof, and the snuff dishes thereof, shall be of pure gold.'

According to chapter 35, verse 28, the rulers 'bring *oil* for the light'. In Leviticus, chapter 24, verse 2, the children of Israel are commanded to bring pure olive oil for the light, to cause the lamps to burn continually.

In the directions for the continual service of the Hebrew altar, minute directions are given for the attention to be afforded the lamps. If the attendant found the two eastern lamps burning when he came to remove the 'ashes', he was to remove the ashes from the others and leave those burning in their place. If he found those eastern lamps extinguished, he was to remove the ashes from them and light them from those burning, afterwards removing the ashes from the remainder. If other than the two eastern lamps were burning, they were extinguished and not relit until the evening, only the eastern lamps being left burning; the candlestick or lamp-stand was obviously placed in an east to west position. In this general act of attention, the old oil, wick, and 'ashes' were removed from the five western lamps: the eastern couple were renewed after the sacrifice.

The direct descendant of the seven-branched candlestick or lamp was the Hanukkah or Chanukkah lamp, but the latter, in its commonest modern forms, is a group of spouted saucers rather than a floating-wick type. It is found wherever the Jewish race have settled and its general 'make-up' is that of a row of eight small spoon-shaped lamps backed by a sconce-like panel decorated according to country and period, and sometimes with a narrow trough as drip-catcher beneath. Sometimes there is also a 'pilot light' in addition, either as an extra lamp in

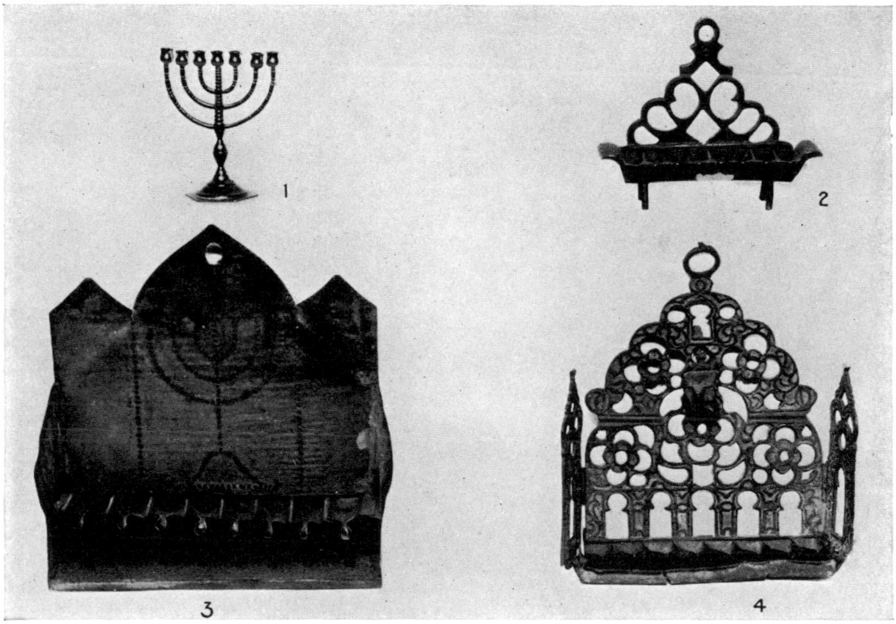

(*a*) HANUKKAH LAMPS, ETC.

(*b*) LAMPS AND TORCH HOLDER FROM MALABAR COAST

the row or separately. One Hanukkah lamp in the possession of the author emphasizes the link with the Hebrew 'candlestick' by an embossed representation of the latter on the back panel.

The origin of the form of the Hanukkah lamp and its possession of eight lamps instead of the seven of the Hebrew 'candlestick' is said to have arisen from the rededication of the temple by Judas Maccabaeus in 165 B.C., after its pollution by the pagan altar set up by Antiochus Epiphanes, king of Syria, three years before. When the perpetual light of the temple was to be relighted then, it was discovered that only one flask of oil, sufficient for one day, remained, but this miraculously lasted for eight days; in commemoration of this, the Hanukkah lamp is lit one burner or lamp at a time on successive days, to cover the span of eight days of the festival period.

There is, however, a striking resemblance between the arrangement of the Hanukkah lamp and that of certain other eastern lamps, particularly Indian temple lamps of the 'goddess' multiple reservoir type (even to the provision of the pilot light in some examples), and it is impossible to avoid suspecting some common connecting link or idea. One Indian lamp in the Smithsonian collection is even nearer to the seven-branched candlestick, having five unspouted bowls on branches attached to a centre pedestal. Mr. Hayward (U.S.A.), in *Colonial Lighting*, records a 'home-made' lamp used by the early Jesuit missionaries among the Indians of the Western Hemisphere, in which the row of eight lamps is employed, in striking resemblance to the Hanukkah lamp, but of course this may be due to the influence of the Holy Land.

[*See Plate XVII.*]

CHAPTER XVI

INDIAN TEMPLE AND VOTIVE LAMPS

THE ordinary peasant or domestic lamps of the Middle East are of the open bowl type, in earthenware or metal, plain or decorated, but nearly always (as distinct from those of the Far East) with a spout or lip of some sort. In the earthenware lamps this is usually a pointed 'beak' or lip with no great projection, but in the metal lamps it ranges from a pointed end to the bowl itself to a projecting trough or open channel sometimes as long as the bowl. They are not handled, though a Kashmir type has a leaf-like appendage at the back rather reminiscent of the attachments of some Roman lamps (No. 6, Plate X, *b*).

The lamps of the temples, shrines, and palaces are mostly open bowl lamps too, but mounted as the useful nucleus in a more or less dominant ornamental whole, naturally—in the case of temple and votive lamps—enshrining elements of religious symbolism. It would be impossible, in a general summary of the lighting appliances of the world and all ages, to cover the whole field of variation in Indian lamps; they are almost endless, so far as differences in detail are concerned. Like Roman lamps, however, while offering continual variations, they can be (with perhaps some exceptions) grouped into certain general outlines, and the author has attempted, in this chapter and in Plate XVIII, to present a representative selection of types.

A very simple form (No. 1, Plate XVIII, *b*) has a small bowl, sometimes round and (unusually in Indian lamps) spoutless—which latter fact raises the suspicion that it may have been intended for burning camphor for ceremonial purposes rather than as a true lamp. There is a specimen of this kind of 'lamp' in the Smithsonian Institution collection with a bowl in the shape of a lotus pod. The bowl is on a short stem and there is a sinuous handle behind which, instead of connecting directly with the base of the bowl or stem, is attached to it by a horizontal bar connecting to the base of the upright. In some cases these lamps have the cobra head as a continuation of the handle (which latter evidently represents the body of the snake) curving over the back of the lamp. The handle enables the worshipper to wave the lamp before the image of the god.

Another small lamp (not here illustrated) has the bowl standing on a low pedestal, the handle being closer to the lamp than in the form just dealt with, and the wick (or wicks, as they are sometimes multiple) resting in a shallow trough channel (or channels, up to three, side by side). Here again the cobra head is over the back of the bowl.

A large number of Indian brass votive lamps take the form of a female figure holding before her the spoon-shaped reservoir or reservoirs (Nos. 2 and 3, Plate XVIII, *b*). In the simplest form of this type it is a single bowl she holds, which, in the smaller versions, could contain oil to burn for a short time only, but there are a number of variations in multiple lamps of this form (known as *dipalakshmi* in allusion to the luck-bringing qualities of the goddess). In some of these the figure carries before her a semicircle of spoon-shaped bowls (sometimes detachable) with perhaps a pilot light or lights above. In another variation she is mounted on an elephant with the little lamp bowls before her, while the animal itself carries the pilot light in its trunk (No. 3). Despite the presence of the female figure, there is a striking and hardly accidental analogy between these multiple lamps and the Jewish Hanukkah lamp, even the number of bowls (eight) often being the same, a rather strange coincidence in view of the fact that the Hanukkah lamp itself is peculiarly a Jewish form and has historical and religious associations which make its symbolism meaningless to non-Jewish races.

The spoon-shaped bowls (from as few as two upwards) also occur in multiple lamps without the figure of the goddess but with the curious arrangement of the handle present in Nos. 1 and 3. In these, the upper end of the handle may be a cobra head, but is often conventionalized into a mere knob (No. 5, Plate XVIII, *b*).

Returning to a simple form, but placed here because it is a multiple lamp on the lines of the *dipalakshmi* row of bowls, there is a type of votive lamp in which a row of small pointed nozzles is fed by a narrow trough at right angles behind. Behind the trough again is a horizontal handle, the whole being supported on three studs (one under the end of the handle and the others at the two ends of the reservoir) (No. 4, Plate XVIII, *b*).

There is a large group of Indian brass lamps which may be called the 'vase' group. There are two forms in it, but both have as their main feature a deep vase-like body, evidently forming a reservoir from which to replenish the lamp bowl itself. At the back of this reservoir there is a snake handle terminating at the upper end in several cobra heads in a group overhanging the back of the vessel and with a small figure of a divinity overshadowed by them. The two forms differ in the arrangement of the lamp bowl itself. In the less elaborate form (No. 6, Plate XVIII, *b*) there is a circular trough surrounding the top of the vase, which leads to a long projecting channel for the wick. In the other (No. 7) the lamp bowl is a separate shallow leaf-like dish before the lip of the vessel and separated from it by a floral ornament. These lamps are usually provided with a spoon for ladling out the oil and are Nepalese in origin.

There is a similar type of lamp from South India in which the vase receptacle has a long flat or nearly flat handle projecting behind and a shallow saucer for the wick projecting in front. In this type the central vase itself is usually much smaller than in the others, in proportion to the size of the lamp.

Most, or all, of the foregoing lamps are for the use of the temple worshipper, who purchases a spoonful of oil with which to keep the flame burning during the offering of his prayers—a notable parallel to the sale of candles for votive purposes in Catholic churches. The lamp bowls are small (mere 'spoons' in the cases of Nos. 2, 3, 4, and 5) and shallow. The lamps illustrated in Nos. 8 and 9, however, are without foot or pedestal and have larger bowls.

No. 8 illustrates a common but sometimes very elaborate form. The general outline is that of a pointed ovoid bowl of fair depth on a hollow throne-like base with a shield-like back having serrated edges. On either side of the lip or nozzle are two projections, and in some cases there are 'pilot lights' of spoon form above the level of the main lamp bowl. The back shield is embossed with considerable ornamentation of a religious or legendary character. A bronze example in the possession of the author (probably older than the more usual brass examples) has a very elaborate design of the goddess Lakshmi and elephants on the back shield, two small pilot lights at the sides, a mounted man at each side of the raised hollow base, and a bird beneath the lip. While this type of lamp is a common Hindu form in everyday use, it is also used by Indian Catholics, in which case the Hindu symbols usually present by way of decoration are replaced by the cross.

The type shown in No. 9 is a modified form of the No. 8 or 'throne' type, in which the bowl is shallower, the base quite flat instead of a raised hollow, and the ornamentation simpler.

Having something in common with the two preceding types in the form of the lamp itself, there may be described here a lamp which appears to be peculiarly Nepalese rather than a general Indian form. In this, the shallow bowl is provided with an openwork back enshrining the figure of a divinity. Instead of the low base, however, the lamp is mounted on a tall, heavy pedestal out of all proportion to the insignificance of the lamp bowl itself. The latter is surrounded by a circle of pendants (No. 3, Plate XVIII, *a*).

None of these lamps have the useful portion of any great size or designed to give a wide range of light. In the next two we have a group of lamps capable of and intended for reasonable illumination.

No. 10, Plate XVIII, *b*, illustrates a form of lamp having a bowl with several fairly large spouts around its circumference. The bowl is situated around the middle or below the top of a central stem which stands in a large dish base and terminates at its upper end in either (*a*) a loop large enough for a hand hold or (*b*) a loop to receive a chain for suspension, as shown in the specimen illustrated. In the standing variety (*a*) the bowl is sometimes made to slide up and down the stem as in the case of the Italian lucerna (a relation of which is actually used on the Malabar coast). These lamps may be a foot or more high, with a wide dish base, and usually have five or more 'spouts'. Bearing on these, Mr. Gravely, of the Madras Museum, writes:

'South Indian pedestal lamps, which you mention in connexion with the hanging lamp shown in figure 10, are quite common up to about 3 feet high, often with a figure—usually a stylized swan—rising from the centre of the bowl. Such lamps are domestic lamps, though specially valued for ceremonial occasions. And there are wedding lamps of this type with

INDIAN BRASS AND BRONZE LAMPS (*a*)

INDIAN BRASS AND BRONZE LAMPS (*b*)

tiers of up to seven branches each, each branch ending in a similar five-wicked lamp, thus giving sometimes about 100 flames to a single pedestal. The figures are ordinarily solid, so not oil receptacles, though closely related to the bird lamps you describe.' (*Infra.*)

A somewhat similar formation of the nozzles to that in fig. 10 is to be seen in a lamp set at the top of a column or pedestal with a convex instead of a concave base. In this case the nozzles or wick channels are shallower, though the pedestal may be higher (No. 2, Plate XVIII, *a*).

Rarely, a gravity lamp can be found amongst these Indian lamps. One in the author's collection has four large spoon-like bowls connecting to a central basin over which is set a lemon-shaped container having a projecting spout. At the base of the container are holes connecting with the basin. The whole arrangement is set on a high pedestal with a convex base (No. 1, Plate XVIII, *a*).

Cingalese Lamps. A gravity lamp is a very typical product of Ceylon. No. 15, Plate XVIII, *b*, the usual type, works on a somewhat similar principle to that of the bird fountain. A baluster-formed reservoir is attached to a base which is unscrewed and the lamp inverted for filling. In use, the oil is supplied to the wick by gravity pressure through a base tube (not an open channel) leading to a splayed burner evidently intended for a thick wick. Single burner specimens are usual, but multiple examples are also in existence.

Bird Lamps. These are also a departure from the usual principle of the simple open saucer in Indian lamps, and are gravity lamps. They are artistic and, in a simple way, scientific. The main reservoir is in the form of a complete bird with hollow body; at the feet of the bird is the lamp bowl or trough—long, narrow, and open. This is detachable from the body of the bird to facilitate the filling of the latter by inversion. When the lamp is in use and the bowl is full, the level of the oil is high enough to close the aperture in the base of the bird, but as soon as this level sinks sufficiently, oil is forced through a spout in the breast of the bird to replenish the bowl beneath. These bird lamps are for hanging. The one illustrated (No. 14, Plate XVIII, *b*) is for a single wick, but the author has seen one with two spouts which was also lacquered in red.

Lotus Lamps. These are ingenious and of some considerable artistic interest. The pure form of the lotus lamp is simply that of a lotus flower set on a cup on a low pedestal. The petals are separate, and in one handsome form (Plate XVIII, *b*, No. 12) are brass in the outer row and copper in the inner. When not in use, the form presented is that of the closed bud, but on unscrewing the bud to use the lamp the fuel receptacle appears as a round bowl forming the centre of an open lotus blossom. In later editions of this lamp it is elaborated by placing the lotus on the back of a buffalo, with the conventional cobra as back strap and hood to the lamp (No. 13, Plate XVIII, *b*). It should be noted that, while this is often classed as a lamp, the absence of a spout suggests that its primary purpose was for burning camphor.

No. 11, Plate XVIII, *b*, appears to be more a domestic lamp than the other types. It is practically the ordinary domestic brass-spouted saucer mounted on a tall pedestal.

All these lamps are in brass (or more rarely in bronze). They do not exhaust the variations of Indian votive lamps but provide a series of types to which it will probably be found that most specimens can be more or less approximated. The rolling lamp (Persia and North India) has been omitted as being more of a lantern type than an ordinary lamp, the lamp within, too, being of a closed form of entirely different type to any of those dealt with above (Nos. 9 and 10, Plate XXVI, *b*).

[*See Plate XVIII.*]

CHAPTER XVII

MEDIEVAL LAMPS

THE story of medieval lighting in Europe is as dim as the lights themselves were. There are a few allusions to lamps in medieval documents (more particularly on the Continent), but very few examples (and these chiefly late) have survived; they are seldom illustrated in the illuminated manuscripts, paintings, or sculptures of the period. Such lamps as existed were undoubtedly of the open type and considerably inferior in scientific principle to the closed lamps of the Romans or of the medieval East.

So far as the so-called 'Dark Ages' were concerned, the Museum at Oslo possesses an example of an early open saucer lamp taken from a Viking ship; this is on a long stem without a base or feet, and appears to be for sticking in something—possibly the floor or a wood block. A rude clay lamp, cup shaped, with a handle and on a stem, was found at Danes' Graves, near Pocklington, Yorkshire, and two plain-bowled pottery lamps on low feet, showing signs of fire, were found on a Saxon site at Broadhay, Winchester, in 1934. More recently (1937) another pottery 'cup' on a stem which appears to have had a splayed out base (missing) has reached the Winchester Museum; the shallow bowl, of small diameter, shows considerable traces of burning, and this too was undoubtedly a lamp, possibly of the Saxon period.

Reference has already been made (in the chapter on 'Float-Wick Lamps') to the Anglo-Saxon glass bowls from Faversham and the pottery bowl (seventh century A.D.) from Bourton-on-the-Water, which appear to be lamps of the float-wick type.

Rather more controversial, but nevertheless to be taken into consideration as possible lamps, are hanging bowls of the 'dark ages' and of Celtic Ireland which might well have carried oil or fat and floating wicks. A bronze example of Jutish origin, from Chesel Down, Isle of Wight, has a design inside which is advanced as evidence against this theory, but the whole matter is one which must remain in doubt until more positive evidence, one way or the other, is forthcoming.

Paul Lacroix, in *The Arts of the Middle Ages*, illustrates two hanging lamps of the ninth century from miniatures in the Bible of Charles the Bald and states that such lamps were made, according to the means of those for whom they were intended, of baked earth, iron, brass, and gold or silver, all more or less ornamented. He adds that 'lamps and candlesticks are not infrequently mentioned in the inventories of the middle ages' and that 'the use of lamps was all but general in the early days of the monarchy'. Even if this is true of France, it does not appear to have been so of England. Instead, it seems certain that in the 'dark ages' (the Saxon and early Norman periods in England) the torch, and in the later days of the Middle Ages the candle, were far commoner means of illumination than any form of lamp.

Where crude earthenware lamps were used, these appear to have been spoutless and of the floating-wick type; the illustrations drawn upon by Lacroix suggest that this was also the case with metal examples. In addition to the pottery lamps to which reference has already been made, a pottery 'cresset' of the twelfth century is in the London Museum, but such earthenware lamps are very scarce—a fact which may, however, be due more to their fragility and low value than to their rarity in the days of their use.

The probability of these lamps of the Dark and Early Middle Ages being float-wick lamps is emphasized, and their probable origin suggested, by illustrations in the illuminated manuscripts of the twelfth and thirteenth centuries showing the Byzantine glass lamp, with floating wick, in use—generally with a very lurid flame which indicates either that more than one wick was in use, that something was on fire and that the lamp would speedily crack (!), or (perhaps the most probable explanation) that the medieval artist had somewhat exaggerated the 'candle power'. Three of these illustrations from the illuminated manuscripts are reproduced in Green's *Short History of the English People*, and it will be noticed that, while the documents in which they appear are ecclesiastical in character, the scenes depicted are in surroundings in which the lamp used would probably be of an ordinary domestic type—e.g. the Nativity, usually shown in the setting of a medieval bedroom, and a scriptorium, which, being one of the ordinary chambers of the monastery, would be lighted in the same way as the domestic apartments. So far as the author is aware, however, there are no representations of such glass lamps in any document later than the thirteenth century, though of course the type of lamp itself has survived in the churches until the present day, just as the pricket candlestick has, though the latter was dismissed from ordinary use by the sixteenth or seventeenth century.

A much more primitive form of lamp survived until at least the close of the Middle Ages, but seems to belong more particularly to the earlier medieval centuries. This was the stone lamp (or cresset stone, in its medieval form), very little removed, except in 'finish', from the hollowed stones of prehistoric lamp-makers. Early examples in Winchester Museum show the cresset lamp in single form, as a cup about 8 inches high with solid handle and as a nearly diamond-shaped hollow stone with a circular depression at one end of the bowl (probably to hold the base of an upright for supporting the wick); a smaller edition of this latter type, in the Guildhall Museum, London, is dated as twelfth century.[1] Another in Winchester Museum is a block of stone having four circular hollows and two solid handles; each of the hollows has a small hole in the centre for the wick support, and this feature is common in medieval

[1] The cup-like lamp in Winchester Museum is similar in form to a pottery cresset from Lincoln, in the British Museum, classed as Romano-British.

cressets. It is present in a grotesque-looking stone vessel in Wells (Somerset) Museum which may be a later example of the stone lamp; this vase-like receptacle shows grooves in the sides as though to hold a transverse support for the wick holder (as is the case with some glass float-wick lamps).

Two more good examples of cresset stones in Hampshire are in Romsey Abbey (one with two hollows about 4 inches across, the other with three of about 3 inches diameter, the stone 'body' in each case about 4 inches high—a common dimension). These also have the central holes to the hollows. A cresset stone at Salisbury with five hollows has not the central holes.

A large proportion of the extant cresset stones are multiple; one at Lewannick (Cornwall) has seven depressions, each about 3 inches in diameter and about 3 to 3½ inches deep. Another, at Newtonhall, Wigtownshire, has five depressions only 1 inch in depth. One at Brecon seems to hold the record with thirty hollows. Others are at Calder, Furness, Waverley, Llanthony, York (St. Mary's), Kensey (Cornwall), Parwich and North Wingfield (Derbyshire), Walton and Weston (Yorkshire).[1]

An object recently discovered at St. Paul's Church, Exeter, behind a seventeenth-century monument, is probably a lamp and is of hand-cut Beer stone, circular, 4 inches diameter, having a central hollow only 1 inch in diameter with a small diagonal piercing (possibly a wick-channel) in the rim, connecting with the bottom of the reservoir. Curiously enough, the nearest parallel to this is a crude hollowed stone said to be of Roman date, from Westbury, Wilts., now in Devizes Museum and described as possibly a lamp; this one is lidded.

An example of a single cresset obtained by the author in Devonshire looked like a stone mortar but was found to be coated with old grease at the rim and half-way down the bowl (the use of water for the grease to float upon, as in modern float-wick lamps, would have kept the lower part of the bowl clean). Another, a square block of red stone with a small projection as handle, must be a very late specimen as it is coated inside with a glaze suggestive of the sixteenth or seventeenth century; the provenance of this is doubtful and it may not be English. Yet another stone, in Devizes Museum, is hollowed out to form a lamp of which the date is uncertain; it is probably medieval.

In the west of England, even where no cresset stone has survived or been traced, several churches have chimneyed niches which are alleged to have contained them. A writer in a west country journal suggested that the church cressets in that part of England served as 'public matchboxes', to provide means of lighting fires when striking a light was a matter of difficulty. This may have been an incidental use, reminiscent of the communal fires of primitive days, but to give such a secondary use the position of the primary one is another example of the process of seeking a complex explanation when a simple one is just as likely to be correct—a process to which some archaeologists seem prone.

[1] Cox, *English Church Fittings*.

Mr. Cox, in ***English Church Fittings***, says: 'Cresset was originally a term applied to a cup of earthenware or metal fastened on top of a pole and used in medieval days as a kind of portable lantern. This cup or receptacle contained oil and a floating wick, and the term was eventually transferred to a stationary light formed out of hollow cups carved in the surface of a stone and hence called cresset stones.' To whatever other uses they may have been put, cressets were certainly used in the monastic houses to light the cloisters and other parts of the conventual establishments after dark, and it was the duty of the kitchen officials to provide fat for and fill the cressets. The sixteenth-century inventory of the *Ancient Monuments of Durham Cathedral* mentions a twelve-holed cresset at each end of the dorter (dormitory) to light the monks when they rose for matins (soon after midnight) or 'for their necessary uses', as well as one or two cressets at the choir door (presumably to light their way to the first service of the new day). All these lamps were fed with tallow, which the Durham inventory specifically mentions was provided by the cook.

Sometimes such cresset stones were fixed. There is one on the wall of the slype at Gloucester Cathedral (a Benedictine monastery) and the remains of another are at the junction of the east and south walls of the cloisters of Inchcolm Abbey.

Lys og Lysstel illustrates three late medieval cresset stones—a square stone with four hollows like the usual English type, a handled single bowl, and a sad-iron shape with a single round hollow; all are now in museums, but two at least come from Norwegian churches.

All these cresset lamps seem to have belonged to ecclesiastical establishments of one kind or another and to have been used where a fairly large light was required; even if the glass float-wick lamps survived during the whole of the Middle Ages they would not seem to have been in common use. What then of the ordinary domestic lighting? There is little doubt but that the dip candle was the common illuminant, but such lamps as there were (more common, no doubt, on the European continent than in England) would have been of the open pan or bowl type. A hanging lamp of A.D. 1000 from Ireland is an open affair with three chains, in bronze, decorated with animals' heads and copper inlay. The ninth-century lamps already referred to as instanced by Lacroix were not dissimilar.

After these there is a paucity of survivals of European lamps of definite medieval dating, but there are lamps of the sixteenth and seventeenth centuries in existence which are so primitive in form and principle as to suggest that their pattern is medieval if the actual lamps are not, and there is always the possibility that some of the lamps themselves may be considerably earlier than the date of their last use. Lamps of the crusie type, though most existing examples are only from 50 to 200 years old, must have come down through the Middle Ages in some parts of Europe (though there is no trace of their use in England other than in Cornwall or near the Scottish border) since their prototypes figure among remains of the Roman era. Apart from these, there are a number of examples of simple pan or bowl lamps for fish oil or tallow, in iron and brass, with either a floating wick or one overhanging the rim of the pan.

Very primitive in type are the iron lamps for fish oil hanging in the Hansa House in Bergen (Norway). These are large iron pans with spouts at the four corners, the smokiness of their light being attested by the expansive cowls or smoke shields above the larger examples. The state in which the house is preserved and the dating of the closing activities of the Hanseatic League give a date for these lamps as late as the sixteenth century, but they may well be earlier. Cod-liver oil was the fuel used.

The several museums at Antwerp (the Steen, the Old Butchery, and the Plantin Museum) show examples of open-pan lamps of rather more advanced types in brass, the form usually stellar, with reservoirs arranged in tiers and with taps for drawing off the oil. These are almost certainly developments of a medieval type of lamp, though apparently dating from the eighteenth century or so. Lamps of similar forms are shown in old illustrations of the sixteenth and seventeenth centuries.[1] Curiously, a somewhat similar stellar form of open lamp in the Pitt Rivers collection at Oxford is from southern Nigeria. The idea may have come via North Africa (as the Smithsonian Institution catalogue a lamp of this type from Tangier) but is more likely to be the result of the Dutch trading influence in West Africa which is apparent in some of the lamps improvised in recent times by the natives of that area.

Thirty specimens of a very crude type of sheet-iron lamp were found in Winchester Cathedral precincts during repair work; these were simple circular shallow pans with straight walls, each lamp having two wick-channel tubes opposite one another and a wire loop over with a small hook for suspension. At first sight these would appear to be the rough output of a local smithy working on primitive lines, but similar lamps (one like the Winchester examples, another four spouted, and a third square in form and four spouted) were found in Bruton church crypt and are said to have been used during the funerals of the Berkeleys, who were buried there from the sixteenth century. Except for the probable addition of the wick-channel tubes they are possibly survivals of the tallow-burning lamps produced by smiths of the Middle Ages.

Very similar lamps in copper, deeper, better made, and provided with swivelled hooks, have emanated from Italy, and the fact that they seem to be to some extent uniform in size with the English examples points to some connexion or perhaps some common origin in a pre-reformation form associated with the Church.

The corrodibility of iron probably accounts for the disappearance or 'scrapping'

[1] A six-pointed bronze lamp of similar form in the Guildhall Museum, London, is given as Roman.

of medieval lamps and, conversely, their disappearance suggests that iron was the principal material used in the construction of most of them, since bronze, copper, or latten (brass) would have lasted longer, and pottery, though breakable, does not utterly disintegrate.

[*See Plates VII, XIX, and frontispiece.*]

PLATE XIX

(*a*) BRASS HANGING LAMP OF RENAISSANCE FORM

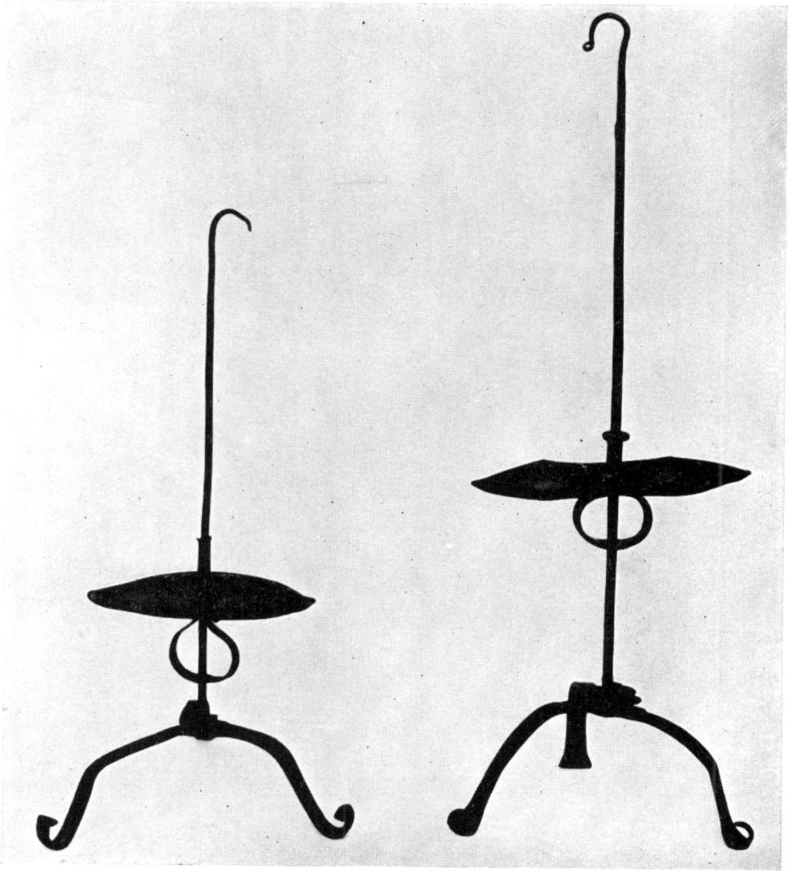

(By the courtesy of the Science Museum, South Kensington)

(*b*) SWISS IRON PAN LAMPS

CHAPTER XVIII

THE CRUSIE AND ITS RELATIONS

THE crusie proper is reputedly Celtic (Early Iron Age) in origin. Whether this be so or not, its original home was probably in the northern parts of Europe and its occurrence in the south may be due to northern incursions. It generally consists of an approximately pear-shaped or ovoid open bowl attached to an upright flat band at the back (the broad end of the bowl), the upright finishing at the top in a swivel connecting it to a combined hook and spike which enables it to be either hung from the roof, the ceiling, or from a special stand, or else suspended from the wall, into which, in that case, the spike is driven, the hook acting incidentally as a support. Most of the typical examples are double—i.e. two bowled, the upper bowl or 'valve' being hooked on to a toothed rack arrangement some little way up the upright, while the lower bowl, permanently attached to the base of the 'strap', serves as a drip catcher. This double arrangement, with drip-catcher, is almost invariably found in the Scotch examples and, if the traditional economical attributes of the Scotsman are to be taken into consideration, it might even be that his ancestors were responsible for the invention! The rack arrangement is really another economy device, enabling the reservoir or top bowl to be tilted to varying angles as the oil in it gets low.

Few examples of the earliest stage of the crusie exist; this is not surprising, as it is generally made of iron and therefore subject to corrosive influences. A single bowl 'crusie' (to use the Scottish name) in the Guildhall Museum (showing the typical hook and spike found in all the later examples) is given as 'Roman' and is probably a British product of the period of the Roman occupation. The same significantly tell-tale hook and spike are shown in some other examples of the Roman era—e.g. two from Pompeii in the Naples Museum and a flat-bottomed lamp of true crusie form in Berne Museum classed as 'Roman'. The hook-and-spike arrangement is the most constant feature of the whole group of lamps deriving from crusie forms and a vital factor in tracing development, origins, and distribution, especially as, while preserving the same general outline, there are slight local differences in the hook which are valuable clues to the actual country from which any particular form or specimen comes.

Most existing examples of the crusie form of lamp are not likely to be more than say a couple of hundred years old. This leaves a big gap between the period suggested as that of their origin and the dates of specimens which can be used as definite evidence of their distribution. The facts that there is a passing resemblance to the 'valve' form of sea-shell in many of the examples and that the crusie type is particularly prevalent in sea lands (side by side with the sea-shell, for instance, in the Orkneys and Shetlands and in Cornwall) suggest that the type has its origin in the shell lamp, just as did the Phoenician pottery lamp. The persistence of the primitive form of the open lamp until recent times, too, is evidence that it has come down through the ages almost untouched in the areas in which it has been found. A strange feature is the lack of conclusive evidence (so far as the author can ascertain) of its use in England, other than in Cornwall and possibly the parts nearest to Scotland, though the Abbey Folk Museum at New Barnet has a lamp with a single crusie-like bowl, but with a horizontal handle, which was found at York associated with items of the fourteenth century. That it was in England in immediately post-medieval times has been suggested by other writers, but those who have adopted this idea may have been misled by the exhibition of such lamps in collections of English Bygones in certain museums; the writer has made inquiries in a number of these cases and has found, in each instance, that the provenance of the lamp in question was either the Scottish islands or unknown. Medieval England must surely have had lamps of some such type. What and where are they?

The author has not been able to trace a specimen of definitely Welsh provenance, but Mr. G. E. Evans, of the Carmarthen Antiquarian Society, assures him that one was illustrated in the Society's 1911–12 Transactions and that his father had seen such lamps in use in Carmarthenshire, *c.* 1820–30.

In the Orkneys and Shetland, the Hebrides and the islands of the north and west of Ireland, it has only died out recently (in fact it has not entirely died out yet in such places as the Isle of Aran). In these quarters it was generally used with fish oil as the fuel; experiments the author has made with such a fuel suggest a very strong reason for the islanders living outdoor lives and going to bed early! On the mainland of Scotland, where the crusie was also in use (at any rate in the eighteenth century and into the nineteenth to some extent) it seems to have been fed sometimes with tallow, which (while burning at any rate) is considerably less offensive.

In the Channel Islands a precisely similar lamp was called a crâsset or croiset—a word of apparently similar origin to 'crusie'—and here too it has only died out within comparatively recent times. As is to be expected, it is to be found also among the Bretons. Mr. John Hocart, in a paper read before the Guernsey Society of Natural Science in 1908, gives some interesting sidelights on the crâssets, which he says were still in use thirty years before and only finally died out on account of the difficulty of getting cheap oil after the paraffin lamp came into fashion; fish oil was generally used in the crâssets and, not being purified, gave a yellow smoky light and a marked smell. (The author, having experimented with *purified* cod-liver oil, can vouch for the smoke and smell even with that.) The oil vessel was kept in a niche or hole near the oven, concealed by a curtain. The wick was a cotton rag, twisted and singed to keep it together, and was trimmed with a piece of wire, an old nail, or a small piece of wood kept in the crâsset and known as *l'amichet*. The lamp was usually suspended from a lath hanging from the rafters over the long table in the kitchen or the spike was stuck in a hole in the wall. In very cold weather, when the family gathered around the fire, a stand (*le villiain*) was placed beside the hearth and the crâsset hung on it; this was sometimes upset, owing to the unevenness of the clay floors. The place of the crâsset in domestic life gave rise to a number of sayings. Those who complained of not getting their share of the light were told that the crâsset turned its back on the handsomest! An unusually dim light was met with the remark that 'the crâsset thinks we are having tripe for supper'—an allusion to the undesirability of examining the cleanliness of that dish too closely! The term 'crâstâire', meaning originally the amount of oil needed to fill the crâsset, came to be applied to a drink of beer or cider by association with a story of Russian soldiers quartered in the islands entering the cottages and drinking the oil from the lamps. Superstitions, too, gathered around the crâsset. If its flame burnt blue, there would be wind the next day, if green there were witches about, a scab on the wick meant an invitation to a funeral, and a spark flying when the lamp was trimmed or snuffed indicated news for the person towards whom it flew. The Guille-Allès Museum in Guernsey shows some interesting local variations from the normal type of crâsset, two specimens having boat-shaped valves, the lips prow-formed, and another having very shallow bowls (or valves); these are all double. Another, approximating to certain continental developments, is in tin-plate with an iron-wire stem, a simple hook, hinged lid, and square-ended lip, with an upright wick channel inside the lip.

In Cornwall there were two forms of the 'crusie' lamp. One in Penzance Museum is of the normal crusie form, with two pans, each single spouted; but another in the possession of Mr. Gibson, of St. Mary's, Scilly Isles, is a square type with spouts at the four corners.[1] The Cornish type, however, has so completely died out that few Cornishmen know it or even its local name, 'chill' (also applied to an earthenware lamp of different type). An interesting point about its use in Cornwall is that the sea-shell was also used as a lamp there and that one kind of shell so used was that of the mussel. The fact that the shape of the normal 'crusie' valve is not very far removed from that of the mussel shell seems to provide evidence for the shell origin of the crusie lamp, especially as the shell lamp and the crusie type have existed, concurrently or consecutively, in other areas as well.

The true crusie, generally in its double form, was widespread on the continent of Europe, especially in western Europe from Scandinavia to Spain. Continental and Channel Island specimens seem to differ from the Scottish crusies, the back straps of the upper valves (the lamps proper) in the latter usually finishing in volutes or pierced terminals, whereas those in continental forms are usually plain or merely splayed at the top. The shaft of the hook-and-spike of the Scottish variety, too, is often plain while the continental forms almost invariably have twisted stems (occasionally to be found in some of the Scottish examples as well);

[1] The author had a similar four-spouted form in sheet iron, recently made, from the Greek island of Mytilene, given him by Mme Hagimichali, of Athens.

in this respect the Cornish and Channel Island specimens follow the Continent.

Occasionally, the crusie is lidded. The earlier examples and most of the Scottish and Jersey crusie lamps are in iron—in Scotland they were sometimes hammered out in stone moulds, a good example of the latter being in the collection at Marischal College, Aberdeen. Rarely a specimen occurs in a non-ferrous metal. In other countries, however, there was a frequent use of brass (notably in the Low Countries, France, and Iceland).

The single form (without drip-catcher) occurs in Germany, Switzerland, Hungary, Italy, Greece, &c.—occasionally farther north. Sometimes the lamps of this (single) type were suspended in pairs from the arms of a cruciform stand; generally the single lamps were smaller but better made than the double. In Greece the single lamp was suspended from a baluster-like wooden upright, somewhat similar to a crusie stand sometimes used in Scotland. Italian varieties are single, very flat-bottomed, smaller than the true crusie type, and decorated with chased or incised ornamentation. Iron was the material generally used for such lamps in central Europe, but copper, brass, and tin occur in the southern countries. The author purchased two examples in corrugated tin from ordinary tinsmiths' shops in Athens as recently as June 1938, from which it would appear that they are still in use; these lamps had suspension hooks of iron wire only, but the better made Greek specimen illustrated in the frontispiece (which is of tinned iron and decorated) has the typical hook and spike, as has also the four-spouted form from Mytilene referred to above.

Most of the crusie-type lamps betray their common origin with the true crusie through the hook and spike, though the hook, 'fish-hook' form in Scotland, is 'S' form in Spain, angular (nearly square) in Italy and Greece, and angular or heavy and narrow in France. Some of the single varieties have a simple hook without the spike, and occasionally this is to be found in Scottish and other double forms; the combination of hook and spike is more usual, however, from Scandinavia to Greece.

In the Teutonic areas—especially Germany, Austria, and Switzerland—the crude hanging lamp took a form which is sometimes classed separately as a 'pan lamp'. This has a shallow reservoir—little more than a flat surface with upturned edges—and is less standardized in shape than the crusie, varying from square, through heart, diamond, clover-leaf, and triangular shapes to 'three-way' pans. Usually pan lamps are single, without drip-catchers. This fact and the shallow depth would render them unsuitable for burning fish and whale oils, and it is in the inland areas, where the less fluid tallow was the fuel most used, that they are commonly found. Despite the difference of fuel and bowl, these pan lamps are essentially close relations of the crusie and are adaptations of that vessel to tallow burning; their relationship shows not only in the general design and principle but also in the frequent possession of the crusie hook and spike.

Closely allied to this 'pan' type of hanging lamp, but otherwise showing no

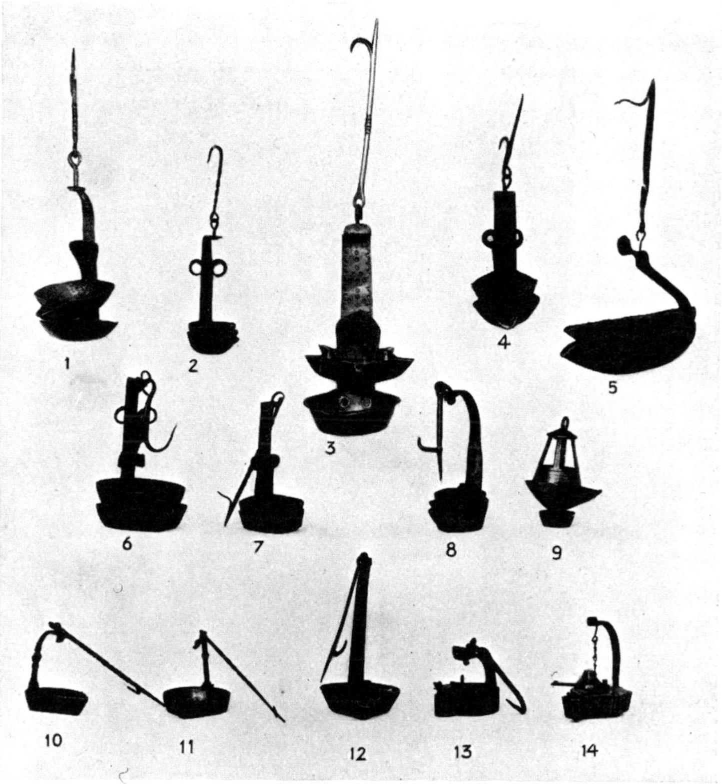

(*a*) CRUSIE TYPES

(*By the courtesy of the Neprajzi Museum, Budapest*

(*b*) OPEN IRON AND COPPER LAMPS, HUNGARY

relationship to the crusie types, are the standing 'pan' lamps of German-Swiss areas, in which the same pan arrangement for burning a wick resting in tallow is attached to an upright on three feet—either one pan surrounding the central upright or a pan on each side of the upright. In all the pan types, economy in the filling of the reservoir was sometimes effected by the partitioning off of a part of the pan nearest the wick by a low barrier. According to Dr. Rushford (*Hobbies*, September 1937), the pan lamps of Pennsylvania were provided with 'pushers' to keep a supply of the semi-solid fuel near the wick.

The crusie was barbaric in its simplicity and behind the Roman lamp in scientific principle. In fact, it was an absolute reversion to the open lamp of the most primitive cultures. In continental forms, however, improvements gradually crept in. The first was a tongue or tube in which the wick was laid or inserted, reputedly first introduced in Italy but present also in Dutch, Spanish, and Greek forms (see Nos. 8 and 11, frontispiece, and No. 11, Plate XX, *a*). Later, a hinged lid or fixed top was added; a sheet-iron form prevalent on the borders of France and Italy has the fixed top with a small hinged lid let in (No. 12, Plate XX, *a*). Another version presents the appearance of a padlock, with a sliding arrangement to cover the filling hole; this form hails from several countries in south-west and central Europe but is often classed as a mine lamp. It is to be found both in iron and brass (Nos. 13 and 14, Plate XX, *a*). None of these later developments nor the substitution of tin and brass in many of the later covered forms absolutely displaced the primitive open iron form, which has persisted in several outlying parts (such as the Scottish and Irish islands) until recent times and in a few remote corners even to the present day.

Other variations and 'relations' of the crusie are numerous. France, especially, possesses a number of types based on but diverging from the simple crusie. One French lamp, in copper and iron, in the Smithsonian Institution collection, has the lamp reservoir suspended by a curved ratchet 'handle' from a separate pendant, instead of the ratcher being on the upright. Another, in iron, is a hand lamp of crusie form with a flat handle at the back; this is said to be of the fourteenth century (cf. example from York in the New Barnet Folk Museum quoted above). From France, too, comes an elegant brass crusie with a decorated upright and five spouts to the lamp saucer—the drip pan underneath being a plain round affair devoid of a spout—in the author's collection (No. 3, Plate XX, *a*); a similar lamp in the Smithsonian collection is associated with Flanders and there is one in the Berne Museum. (These brass multiple-burner hanging lamps have very close counterparts in open spouted hanging lamps from Morocco, which, however, do not carry the crusie form of hook and spike and are probably of independent origin, though very near in general design.)

A more common form in France is a type from Normandy, in brass, with a pear-shaped reservoir fitted above with a bell-shaped dome or 'annexe' connected to it by narrow bands, a small drip catcher being suspended beneath the reservoir by a hook and loop. An open side projection provides for filling. A specimen in the author's collection (though this was stated to be of Dutch manufacture) was dated as sixteenth or seventeenth century, and there is little reason to doubt their having been in use at that period or even earlier (No. 9, Plate XX, *a*).

The enclosed or lidded crusies lead on to enclosed types which obviously derive from the crusies and, while showing little or no relationship in general form, retain the typical hook and spike. The nearest of these to the primitive crusie types are round to pear-shaped, in iron, tin, or brass, with formed spout, flat-based and flat-topped in body, having an overhanging back-strap to which the hook-spike (or a simple hook) is attached. This group seems to belong more particularly to central and southern Europe and may represent an amalgamation of the crusie with the closed classical lamp. At a later stage the spout becomes more pronounced and projecting and the crusie hook and spike is found associated with globular lamps of developed forms and with definitely projecting and upturned nozzles or burners, many of them provided with feet for standing as an alternative to suspension. Most of these forms are in brass and come from Latin countries, but a pewter form in the author's collection was procured from Amsterdam and has a Dutch appearance, though the hook is of the French type (No. 3, Plate XXII, *b*). These lamps, however, while showing their descent in the presence of the crusie hook, are far removed from the crusie proper and belong rather to the class of spout lamps prevalent in the eighteenth century and immediately preceding the central burner types of fairly recent times.

In the United States of America most of the crusie types and relations, but more especially the pan and lidded or enclosed forms, have been used and a transatlantic nomenclature evolved, the true crusie types being known as 'Phoebe' lamps and the types having a wick support or channel as 'Betty' lamps, the latter term, however, being sometimes loosely applied to all the crusie forms. It seems to be assumed there (in America) by many writers that the Phoebe lamps, at any rate, were imported by English settlers; the absence of evidence of the use of crusies in England rather discounts this, and while Scottish and Irish settlers are likely to have imported them, the particular forms traced in America suggest continental rather than British origins. In Pennsylvania in particular the pan type seems to have been prevalent, suggesting its introduction by German colonists. Having drawn the idea itself from their home countries, the North American colonists proceeded to develop local variations in form and material, most of which betray their origin but add the touches of individual ingenuity and the inevitable exchange of ideas. Several different forms, in tin and iron, are illustrated in Hayward's *Colonial Lighting* and in 'The Lamps of Colonial America' (Watkins, in *Antiques*, October 1937).

[*See Plate XX and frontispiece.*]

CHAPTER XIX

OPEN STAND LAMPS

A PRIMITIVE form of lamp which has survived in many countries in Europe until quite recent times is the more or less open bowl, reminiscent of the very earliest 'lipped' saucer lamps, set on a candlestick-like stem with a dish base.

Cornwall, the home of so much that is archaic and has failed to survive (or has not even existed) in the rest of England, possessed such a lamp in pottery known, like the Cornish form of the crusie lamp, as a 'chill'; probably the pottery lamp was the one more truly entitled to the name. There are simple specimens of this lamp in Penzance and Truro museums; its bowl or reservoir is a plain open saucer with a small spout or lip, set on a short stem terminating at the base in a shallow circular dish, a handle being attached to the stem. This lamp was in use in remote parts of the county as recently as 1860 but has now so completely disappeared that specimens outside museums are difficult to find. The one in Penzance Museum (used at Zennor up to 1860) is about 8 inches high, with a reservoir about 4 inches in diameter, in red-brown glaze; that in Truro is rather shorter and in a greenish-brown glaze, the drip basin at the foot having a spout at the side (as is the case with Swedish examples cited hereafter). A Cornishman once informed the author that these lamps were used with two kinds of wick—a fine wick of reed or a coaser textile one; the latter was called a 'booba'—from which it became customary to call any coarse or clumsy person a booba! Is this connected with 'booby'? So is the history of the lamp wrapped up with the life of the people.

The Smithsonian Institution catalogue a Virginian example of this type of lamp as of English origin; if so, it must owe its introduction, one thinks, to a Cornishman.

From whence did this lamp, peculiarly local in England, come? It does not appear to be indigenous, as there is no trace of its parent, the saucer lamp, in that area, nor is it a lamp associated with Celtic cultures. There is a suggestion of parentage perhaps in Saxon pottery lamps in Winchester museum, but these have neither spout nor dish base. It is possible that such a lamp might have had its ultimate origin in the dim and distant past of the Mediterranean area, but its actual recent form connects it more directly with north-west and central Europe. Metivier, a Channel Islander, in a poem 'La fin du chapitre', refers to a crâsset of clay which was probably of the same type as the Cornish pottery chill, but no example or even picture of this seems to have survived—nor do Breton survivals point to an origin in Celtic France.

A very similar pottery lamp was common among the peasantry of Sweden, and the Northern Museum of Antiquities at Stockholm possesses a large number of specimens from various provinces, mostly in brown glazed ware; they are generally shorter in stem than the Cornish pottery chill. It does not seem to have been common in Norway; no specimen of such a lamp is illustrated in *Lys og Lysstel*. Curiously, two lands suggesting few cultural affinities with Scandinavia, Hungary and Croatia,[1] had the same lamp among the peasantry (it is essentially a peasant lamp); that this is not mere coincidence is shown by the fact that the Scandinavian wooden candlestick also appears in Croatia. Berne Museum also shows examples of such open lamps in brown and green glazed wares and the type seems to range from the north-west to the Balkans generally, though such German forms related to this type as are known to the author seem to be more advanced, with a wick channel or completely closed in. Holland has the type in its simple form (e.g. an example in Delft Museum), but pottery lamps are not common in the Netherlands. A specimen with a wick channel in the Smithsonian Institution collection at Washington is described as English, but if the ascription is correct it was probably either an importation or an isolated case of a copied continental form. Most of these open pottery stand lamps, particularly the Swedish and Hungarian examples, have a spout or lip to the drip basin at the base of the pedestal as well as to the lamp reservoir itself—the lower spout usually set at right angles to the handle and upper spout.

Side by side with these pottery forms, there are occasionally found (e.g. in Sweden and Hungary) analogous lamps in tin or sheet iron, plain or corrugated and with or without a wick tongue. Both the pottery and metal forms are generally described as 'train oil' lamps, but it seems reasonable to suppose that they were in use before (probably long before) the general introduction of whale oil and were previously used with tallow.

Holland, with her love for brass utensils of all kinds, translated the open-bowl stand lamp into that metal, replacing the handle of the pottery lamp with a loop for hanging, at the back of the reservoir—a device which is carried on in the enclosed standing spout lamps of the Low Countries. These Dutch brass lamps are more typical of the Netherlands than the pottery specimens, and show a process of development in that the wick tongue or channel appears in some of them and occasionally one is fitted with a hinged lid—a half-way house to the closed lamp. The useful adjunct of the spout to the lower or drip bowl is absent, however. Of three such brass lamps in the author's collection, showing the development from the plain bowl to the wick-grooved lidded type, the simplest was said to have come from France. It is quite possible that this was correct and that the form was not confined to the Netherlands (see Plate XXI, *b*).

It is curious how, while the actual stages of development of lamp types largely coincide in both areas, there is a sharp dividing line in design as between Europe north of the Pyrenees–Alps–Balkans line and south of it, not merely in one stage but in nearly all. In north and central Europe several types of lamp are

[1] It does not seem to be known farther south in Yugoslavia.

(By the courtesy of the Neprajzi Museum, Budapest)

(*a*) OPEN GLASS AND POTTERY LAMPS, HUNGARY

(*b*) PRIMITIVE POTTERY AND BRASS PEDESTAL LAMPS

more or less common to the whole area; in the Mediterranean basin, Oriental, African, and classical influences combine to effect a divergence, so that in most of the 'stage' groups in the history of the lamp there is a north European form and a south European, save that in some cases north European influences creep over the boundary into northern Italy and Spain, to make exceptions.

This is markedly the case with the open-bowl stand lamp. The Mediterranean forms have features which pronounce them to have had their roots in the remote antiquity of Mediterranean and Near Eastern cultures and render them distinct from the more northern types. A Sicilian type is very near the northern version in form (in a different pottery—usually pink or buff ware overlaid with a whitish glaze), but the spout to the drip pan is missing and the bowl itself bears a striking resemblance to the old 'cocked hat' or shell form of lamp (or more particularly to the deeper, flat-based successor to it). (No. 8, Plate XXI, *b*.)

Even closer to the Phoenician shell-like lamps is the Neapolitan type in widespread use in southern Italy and extending to Malta. In this, the reservoir is distinctly a double form of the ancient open pinched-lip lamp unaltered (which double form, incidentally, appears in early Palestinian lamps of this type—see No. 14, Plate IX). The only way in which this lamp differs from its ancient prototypes (and obvious ancestors) is that it is set on a stand and base and that the pottery is glazed. Tradition dies hard in rustic areas such as the southern extremities of Italy, and in this lamp we have the curious spectacle of a lamp form of some hundreds of years B.C. surviving the much more scientific forms of the Roman era and even modern lamps. In these two-spouted Neapolitan lamps, a handle beneath the flame would of course be inconvenient, so there are two handles set at right angles to the spouts (Nos. 1 and 3, Plate XXI, *b*). The lamps range up to a foot or so in height (as against the comparatively dwarf northern forms), but there is a single spouted variation, with the handle behind, which is shorter than the majority of the double-spouted lamps. All these Neapolitan lamps are white glazed and decorated with polychrome ornamentation of a conventional and simple design. It is noticeable that the southern types of pottery stand lamps generally have deeper and narrower bowls than the northern, suggesting the respective origins in the shell and saucer lamps—in fact there can be no doubt that the Mediterranean forms originated with the mounting of primitive hand lamps on pedestals, probably the result, in the first instance, of an amalgamation of the hand lamp with the previously separate stand on which it was placed. (A similar history, with a different form of hand lamp, may of course lie behind the northern stand types, especially as the crusie lamps, for example, frequently had special stands made for them.)

A third form of these stand lamps in southern Italy should perhaps be more correctly classed as a spout lamp, but it has marked affinities with the open types and, like them, shows signs of being an ancient hand lamp mounted on a pedestal. It is the white glazed butt pottery Apulian lamp already described in a previous chapter (globular, spouted, closed-in form with a funnel at the top for filling) mounted on a column with a shallow dish base. One variety has a tall pedestal and the base is provided with a single side spout like the northern lamps, but the pedestal in others is hardly more than a tall foot. The drip-basin spout connexion with the northern types, on the tall specimens, is rather curious, as, otherwise, the lamp has distinct suggestions of North African influence and the North African lamps do not possess such an arrangement.

A somewhat related form of lamp from Greece,[1] in a black glazed pottery hardly distinguishable from ancient Greek ware, has a globular spouted reservoir completely closed in save for a filling hole on top (No. 7, frontispiece).

There remains another and rarer version of the pottery stand lamp from Italy, having a globular reservoir and twin spouts (opposite each other as in the other Italian lamps of this group) and taking the form of a human figure on a flat base, holding the reservoir on its head, sometimes with the upraised arms acting as handles. A female figure of this type in the Smithsonian Institution collection is given as from the Netherlands, and one in the form of a soldier in the author's collection might possibly have come from that quarter, but the type in general is distinctly and certainly Italian. A second specimen in the author's collection, in which the figure is mounted on a mythical animal, is almost certainly Italian, with polychrome decoration on a white glaze suggestive of the Neapolitan lamps, though the reservoir is the same in shape as the two lamps to which reference has just been made (Nos. 2 and 4, Plate XXI, *b*). There are definitely Italian specimens in Birmingham Museum.

It is not without significance that this same idea of a human figure acting as a pedestal for the lamp bowl occurs in a Minoan lamp (with open bowl) in the British Museum—a fact which seems to confirm the assumption of a southern and ancient origin for it, though the intervening links may at present be missing. It is interesting to note that, though the shell-form reservoir does not occur in these particular lamps, there is still, through Crete, the suggestion of a remote Phoenician influence.

Moroccan influence may be responsible for a white-glazed brown-slip decorated lamp from Spain which the author purchased in Gibraltar; in this the reservoir is definitely a closed one, with a long projecting spout. Apart from the form of the lamp itself, however, its place is distinctly with the open-bowl stand types rather than with the spouted types, though the foot is splayed and not in dish form. (Plate XXI, *b*, No. 6.)

Crossing the Mediterranean Sea to its southern shores, we again find the pottery stand lamp in a third group. In Algeria and Morocco (particularly the latter) these lamps take the form of a partially closed (somewhat boat-shaped) reservoir of small capacity with a pronounced (open) spout. The boat-form bowl type is found in polychrome pottery (No. 11, Plate XXI, *b*) or in green

[1] But without a filling funnel.

glazed ware, with a flat or slightly dished base in the taller versions and a deep base bowl in the shorter (No. 5, Plate XXI, *b*). An interesting and artistic variant had a midway collar reminiscent of seventeenth-century candlesticks and probably derived from the candlestick since Moroccan candlesticks often show it (No. 13, Plate XXIV, *b*).[1]

Another type, in brown glazed ware, has a partially bridged open reservoir and a dish base (from Morocco); this again is the hand lamp, of the Arab type, mounted on a stand (No. 7, Plate XXI, *b*).

These North African lamps are single spouted, with a handle at the back extending from bowl to base. They vary considerably in the height of the column and the depth of the bowl or base, some of the base bowls nearly reaching to the underside of the reservoir.

Perhaps this is the place to mention the Arab or North African lamps of alabaster-like stone, somewhat in the form of a small arm-chair, with the reservoir (shallow) as the seat; some of these date from the Middle Ages and are inscribed with Moslem texts. There is a faint suggestion of this type in the Kabyle lamp (No. 11, Plate XXIV, *b*), but in that case the 'lap' or seat portion is a drip bowl only, with a row of lamps above.

The southern lamps were used with olive oil; the northern types with tallow or whale oil.

[*See Plates XXI, XXIV, and frontispiece.*]

[1] The Abbey Folk Museum, New Barnet, has a fine example of this type of standing lamp, in brass, 4 feet high, with drip bowl, for mosque use.

CHAPTER XX

SPOUT LAMPS

In the course of the development of the open lamp, improved provision for the wick had been effected by the introduction of a 'wick channel', at first half round and later completely tubular—a feature which was sometimes retained even in the closed forms of early European lamps. Side by side with this or independently of it had come a gradual lengthening of the spout projection until it became, instead of a 'lip', a 'nose'. As soon as the projection reached that stage, the lamp entered on a new phase in its history, the connexion of which with the 'crusie' or open-lamp stage was in some cases still, but *only*, betrayed by the retention of the traditional hook and spike.

Generally speaking, the enclosed lamps with markedly projecting spouts may be separated into two large groups (though with cross influences between them) which may conveniently be dubbed the 'South European' and 'North European' types, the line of division being roughly that between the north and south types of the old open pottery lamps except that the latter, taking the two groups together, are more widespread than the spouted types. In dealing with the latter, the author does not include the pottery lamp of the Mediterranean area having more or less of a spout but an open top (as in Spanish examples) because, while strictly speaking spout lamps, their relationship is closer to the open-bowled pottery lamps of Italy and North Africa.

South European. These are, in the main, more or less of 'teapot' or globular form and a general feature is that they practically all have a curved upturned spout, at the end of which the 'burner' is set in a miniature 'saucer' arrangement forming a concave collar around it to intercept the drips from the wick (though too small to hold much and with no provision for draining away). This form of nozzle is not peculiar to the south of Europe; it occurs particularly also in Holland. It is, however, more decidedly a feature of the southern types, in which it is all but universal (if not entirely so) and its occurrence in the Netherlands may be a relic of Spanish influence surviving through the centuries.

The smaller examples of this group of lamps usually stand on a circular foot and have a strap at the back to which a hook (or hook and spike as in the crusie-type lamps) is attached. They are almost invariably in brass, and are single or double spouted; they are provided with a hinged or removable cover.

Probably the best-known type of the southern spout lamps, however, is the 'lucerna' prevalent in Italy (especially from Rome northwards) and also used in Spain, Portugal, and Greece. This consists of a round and lidded body which can be raised or lowered on a central stem passing through its centre. The top end of the stem terminates in a loop handle and the bottom has a circular

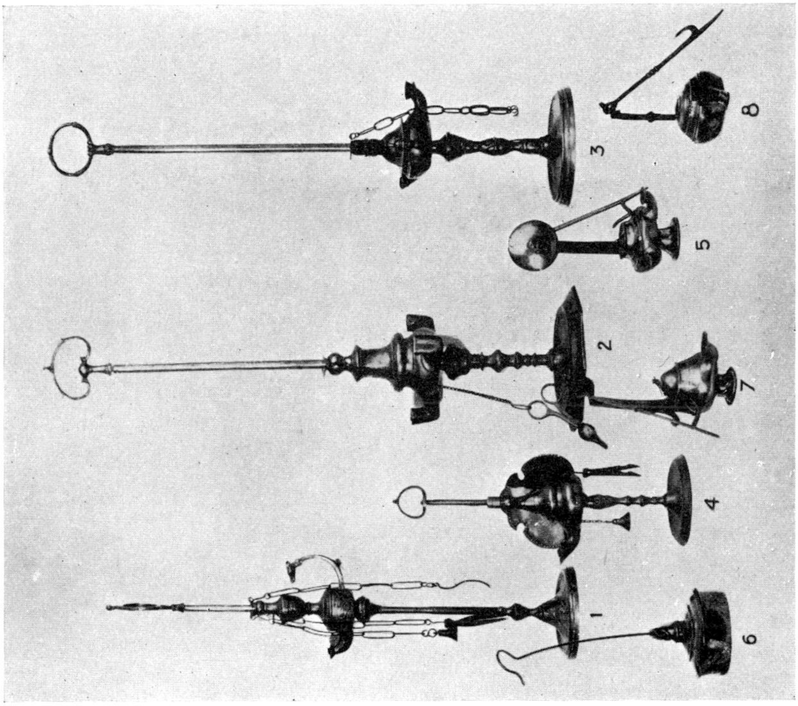

(*a*) SOUTH EUROPEAN SPOUT LAMPS

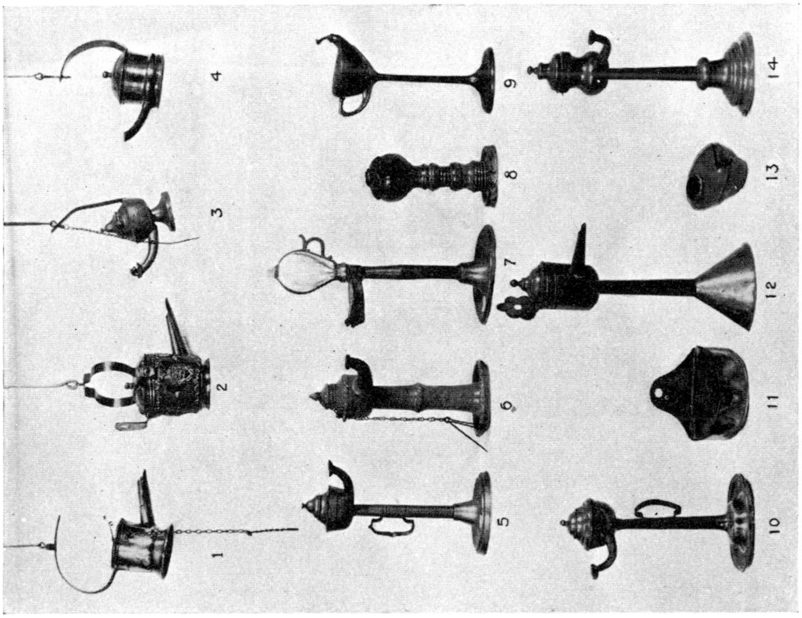

(*b*) NORTH EUROPEAN SPOUT LAMPS

(or sometimes octagonal) base. The common form is in brass. Generally there are three or four spouts, occasionally two (with a curved 'half' handle at the back strongly reminiscent of the handles of some bronze Roman lamps and probably derived from that source); still more rarely, the lucerna is single spouted and the author has a graceful specimen with a movable brass shield reflector at the back (No. 4, Plate XXII, *a*). Another specimen is chased all over (No. 1). Specimens seen include lucerne with small figures of animals as decoration on lamp or pediment. These are olive oil lamps (like the other southern types), are still used in some parts, and are provided with trimmers, prickers for raising the wicks, and tweezers. They range in height, generally, from about a foot to two feet, from top to base. The lamp itself, in its normal position, is usually about midway, despite which it has been suggested that the origin of the standing lucerna lies in the combination of the lamp and lamp stand of the Roman era. Certainly its sliding arrangement was anticipated by a telescopic Roman candelabrum which enabled the height of the lamp it carried to be adjusted, and a three-branched Roman lamp found by the Egypt Exploration Society at Armant is suggestive of the body of the lucerna.

North European types. These present a range of variation considerably greater than that of the southern types, and there are closed pedestal lamps with little or no projection to the spout which may be taken as transitional between the open lamps and the spout lamps proper. Some examples show the same form of spout as the southern types but the lucerna is typically southern, despite the presence of an example in the Antwerp Old Butchery said to be 'local'; the fact that it is described as a 'cloister' lamp suggests that its introduction is connected with southern catholicism.

While practically all the southern spout lamps are in brass (except for rare examples in silver and bronze), the northern types vary in material almost as much as in form, brass, pewter, pottery, tin, and (in certain crude types) iron being the principal materials used.

Dutch and Flemish spout lamps are very typical. Usually they are of 'coffee' pot form, in brass, with the reservoir (having a detachable lid) fitting inside an outer cylinder slotted in front to allow the spout of the inner receptacle to project immediately over a drip channel attached and connected to the outer portion. A pierced attachment at the back allows for fastening to a wall, but most of the examples have a pedestal stand, the base of which is semicircular (to allow the lamp to stand or hang close to the wall) and weighted with sand to give it stability (No. 12, Plate XXII, *b*); sometimes the stem is inserted instead in a wood block as in rushlight holders. Examples minus the pedestal but otherwise of similar form (Nos. 1, 2, and 4, Plate XXII, *b*) sometimes have a loop and hook over for suspension and occasionally the crusie hook survives. Some specimens have the suspension hook connected to the loop by means of a sliding arrangement which permits the vessel to be tipped when the oil is low

(No. 1, Plate XXII, *b*), a principle which again is reminiscent of the crusie types. A further variation in these lamps is a semicircular wall type with a hinged lid, a back shield, and a short spout (No. 11, Plate XXII, *b*), a type which connects with the wick channel open lamps of rather similar form.

These lamps, which are either plain or decorated, were used originally with whale oil and later with colza oil, and most of them belong to the eighteenth and early nineteenth centuries.[1] Their smoky habits are eloquently described by a Dutch slang name for them which may be rendered in English as 'dirty nose'! Decoration, when present, consists of stamped or embossed designs (the lion of Flanders especially) and applied copper work.

J. Seymour Lindsay, in *Iron and Brass Implements of the English Home*, illustrates tinned iron versions of these Dutch lamps as being used in England. The adaptation of the form in iron certainly sounds non-Dutch and the author has also been told that lamps of this type were at one time used in Somerset. Proof of the latter statement, however, has not so far been forthcoming and such lamps must have been unusual in this country; the author has never come across any specimens or heard of any other than those illustrated in Mr. Lindsay's book. A lamp on very similar principles, for the hand or for hanging (principally the latter), was in use for working purposes (on farms, Cornish tin-mines, in cellars, warehouses, &c.) in England, especially, it would seem, in Wales and the west, under the name of 'duck' lamp. Is this term possibly a corruption of 'Dutch' lamp and evidence of its origin? Duck lamps, in tinplate or sheet iron, were in the double canister form of the Dutch spout lamps, without a stand or pedestal, and provided with a loop over for hanging or carrying. Hand lamps were single spouted but the hanging specimens were sometimes with two spouts (opposite one another), each provided with the drip channel. The occurrence of similar lamps in Malaysia was probably due to Dutch influence.

The Dutch lamps may be taken as 'type' forms for the generality of North European (and Russian) spout lamps, most of which resemble them fairly closely in principle, though the body of the lamp may vary in form and non-brass examples are more common outside the Low Countries.

In Germany the reservoir was often spheroid or demi-spheroid. One German form, in pewter, has a glass reservoir above the spout, working on a gravity principle and with a gauge standing vertically against the reservoir (which latter is of elliptical section, with the long side vertically): the gauge is marked with the hours of darkness and acts as a timepiece as the oil sinks (No. 7, Plate XXII, *b*). Naturally enough, this also appears in Switzerland, together with a form like a canister swinging on a cradle surmounting a pedestal and 'coffee' pot forms in tin.

Of two pottery spout lamps in the author's collection, one is English and the other may be. The former, which bears incised lettering on its base, 'Francis, 1754', is in yellowish glaze, with a short spout, and presents a curious feature

[1] Their origin seems to be earlier.

in that a loose lid is provided with an oblong central burner slot of more modern form. Though the lid seems to be of the same ware as the lamp, the anachronism of the central wick (on a lamp already provided with a spout) and the poor fit of the lid rather suggest a later adaptation (No. 8, Plate XXII, *b*). The other pottery lamp (No. 6), though sold as English, looks as though it might be continental. This last-named and the metal lamps Nos. 3, 5, 10, and 14 show the southern drip saucer around the nozzle; No. 3, in pewter, is from Amsterdam, the others (No. 5 in pewter, Nos. 10 and 14 in brass) being of unknown provenance but almost certainly continental.

An unusual form which may be classed with these is in Winchester Museum—this was a covered-in boat-shaped vessel with a very narrow oblong section wick tube rising at one end at an angle, with a large filling hole on the flat top surface. It hailed from Burley, in the New Forest, and its rarity in this country might have suggested a continental origin; the author, however, recently purchased a precisely similar lamp in Salisbury (and therefore in close proximity to the New Forest), so that it may be a type which was in local use at some period or other in the recent past (No. 13, Plate XXII, *b*).

Generally speaking, lamps of any type of the pre-paraffin era are comparatively rare in England. The candle and the rushlight held sway in this country before that, and even in this twentieth century there are rural cottage homes where the candle is the sole illuminant, principally because its portability and the facility with which it can be extinguished and relit render it economical in homes where every penny counts. The use of lamps fed with cheap vegetable oil or animal fat never seems to have 'caught on' much in England at any time, except in Cornwall, where seashells, the crusie forms and the pottery 'chill', fed with tallow or fish oil, were used until comparatively recently; in the mines and mine buildings, not only were duck lamps used but even old brown teapots (filled with tallow), which may, when used for lighting purposes, fittingly be classed as spout lamps!

All the lamps dealt with above were in use in the eighteenth century (generally with whale oil in the north) and were without chimneys. Curiously enough, despite the fact that the central burner (and even the Argand burner on which the modern form of oil lamp is based) appeared within the limits of the eighteenth century, the long projecting spout type of open flame lamp lasted well into the nineteenth century on the Continent; the southern type has not altogether disappeared even to-day. Spout lamps are peculiarly European; nothing of the kind appears outside Europe except as the probable result of European influence (the gravity lamp of Ceylon *may* be an exception but probably owes something to European contacts).[1] The many examples in North America are,

[1] The Government Museum at Madras has kindly sent me an interesting photograph of a group of various spout type lamps peculiar to the Malabar coast, but their striking resemblance to the 'lucerna' indicates their South European ancestry—probably through the Portuguese occupation of the sixteenth century at Calicut, through Portuguese traders or through Catholic missions.

naturally, of European origin, Dutch influence predominating (which again suggests that the lamp was developed abroad to a greater extent than in contemporary England). Local adaptations, however, are frequent. One American lamp of fairly obvious Dutch ancestry has its two projecting spouts made of a flat section (instead of round) for burning lard, a fuel which appears to have been peculiar to America. A special type evolved in New England was the 'Cape Cod' lamp, which, having a body of the Dutch form and an upturned spout without drip catcher, was provided with a charcoal brazier underneath to heat the oil and keep it fluid.

Spout types were handy and were adapted for 'rough-and-ready' circumstances —hence their survival as ships' bunker lamps and for use in factories and mines (see Plate XXVII, *b*).

[*Plate XXII and frontispiece.*]

CHAPTER XXI

CENTRAL WICK LAMPS

REFERENCE has already been made to the introduction, in the crusie type of lamp, of a wick channel in the form of a half section of a tube or a grooved tongue; this appeared also in other open forms of metal lamps but not in the open bowl pottery lamps. It was present in some of the brass open bowls of western Europe. Later, it developed into a short tube which was open for part of the way, as in the case of the little tin open lamps of Spain, and finally into a complete tube. The last-named is found, not only in metal lamps of various shapes, but also in open bowl pottery lamps, especially specimens from the Low Countries and North Italy—but again it did not affect the more typically Mediterranean pottery types.

The Smithsonian Institution catalogue a pewter open bowl lamp on a pedestal having such a wick arrangement, which they attribute to England, but it is doubtful if such a thing was indigenous to this country and if it was really used here either the lamp itself or the idea must have been imported. In this connexion, the fact that several spout lamps of eighteenth-century type have come to the author's collection from East Anglia (without exact information as to provenance) suggests that there may have been such material or ideal importation from the Low Countries into that area, though the types concerned have not been utilized to any extent elsewhere in England.

It is curious that, with all the experiments that had been made through the centuries on the forms of lamps and the improvements in methods of holding the wick, the position of the latter remained unchanged, at the edge of the bowl or reservoir (except of course in the case of floating-wick lamps) until comparatively modern times. Despite the obvious advantages of a central burner, especially in standing lamps, the change does not appear to have really come about until almost the close of the eighteenth century, by which time, or soon after, the first nails had already been driven into the coffin of the old open-flame lamp by Argand's long neglected invention (and in that of the oil lamp generally by Murdoch's experiments with gas lighting).

This can be taken as the real starting-point of the modern rapid progress in the development of lighting appliances. Until then they had differed little in principle from the earliest lamps and, until the use of whale oil in the eighteenth century, the fuels had remained substantially the same—vegetable oils in southern Europe and Asia and animal fats or crude fish oil in the north and west of Europe. Mineral oil was still unknown as a fuel, with the possible exception of oil from seepages in the Mesopotamian oil area.

The wick tongue and tube may be taken as leading in a straight line to the spout, but they also connect to the central wick. Two Pennsylvanian Dutch lamps, for instance, in the Smithsonian collection, show the upright central wick adapted to open lamps, but of course the principle is usually closely identified with the closed lamp in various forms and materials. Probably owing to the contact with the whaling industry and the absence of tradition, the United States of America seem to be particularly prolific in varieties of single central or upright wick open-flame lamps, principally in pewter, tin, and brass and generally in more or less canister or cylinder forms on stems and bases akin to those of candlesticks. Occasionally, crude lamps of this group have a cylindrical body placed horizontally, to accommodate more than one burner. It is interesting to note the combination of this type with the Dutch lamp in the native improvisation from West Africa (No. 6, Plate XXIV, *b*).

The Mediterranean area yields examples with globular reservoirs, usually in brass (see Nos. 4 and 7 from Malta, No. 4 complete with extinguisher and tweezers, and No. 8 from Greece, Plate XXIII, *a*).

The Chinese opium-smoker's lamp is of the central wick type, but is provided with a small globe which can be lifted to enable the smoker to light his pipe —thus bringing it almost into the chimneyed class.

Once the single upright tube type of burner was in vogue, the next step was to endeavour to increase the amount of illumination by the multiplication of the burners in group formation. Benjamin Franklin proved that a two-burner lamp of this type would give more light than two single-burner lamps and a large number of lamps were provided with this improvement. An attempt to carry matters further by using three burners in a group proved a failure and three-burner lamps remained abortive curiosities.

The use of the upright tube burner was associated with a considerable use of glass lamps. At first these had the burner or burners inserted in the top of the reservoir in a cork, but later threaded collars of pewter, in which the burners were set, were fixed on with cement. These glass lamps were generally small 'pedestal' lamps and seem to have been more often adopted in England than the tin, pewter, and brass varieties, judging from the number of lamps of the respective materials extant which can be safely attributed to this country. Most of them carried two burners, but few of the examples now obtainable have the burners with them at all.

Usually these central wick lamps were used, in the first instance at any rate, with whale oil. In the United States, if not elsewhere, the increasing price of this oil eventually led to a widespread use of lard oil. About 1830 the search for a better fuel resulted in the introduction of camphene—a mixture of turpentine and alcohol. It lasted about twenty years only and never became very popular, its principal disadvantages being its volatile nature, which made precautions against evaporation necessary, and its explosiveness. While many camphene lamps were of the single-burner type, they were quite commonly two-burner lamps and in this case the arrangement of the burners was very distinctive in being long tubes inclined away from each other, each tube being capped

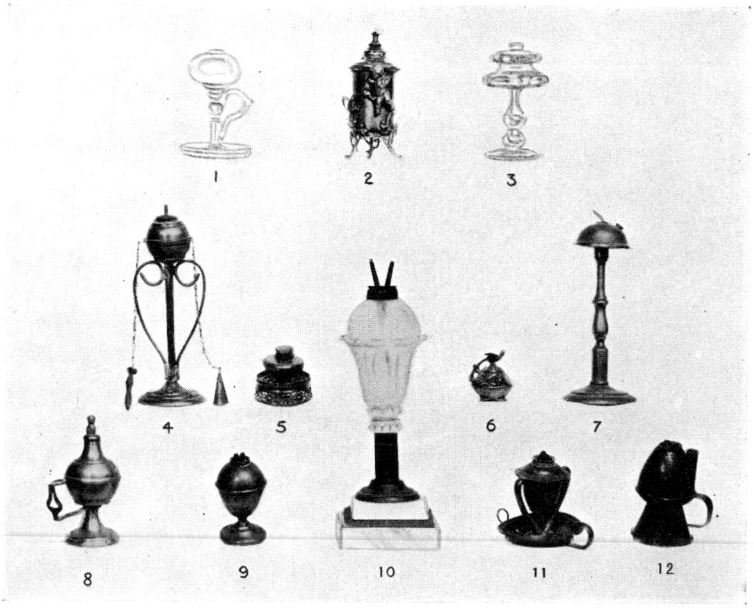

(*a*) CENTRAL BURNER LAMPS

(*By the courtesy of Mr. Julius Daniels*)

(*b*) CENTRAL BURNER LAMPS FROM THE U.S.A.

when not in use, to prevent evaporation from the wick (No. 10, Plate XXIII, *a*). Improvements on the ordinary form of lamp included a gravity type on 'bird fountain' lines which, in addition to assisting the supply of oil to the wick, ensured that the flame was removed some little distance from the fuel reservoir.

Benzene, first discovered in 1825, was also used as an illuminant in the middle of the nineteenth century, usually in small lamps of the central wick type.

All these open flame lamps had solid round wicks of small diameters.

[*See Plate XXIII.*]

CHAPTER XXII

THE MODERN OIL LAMP

THE distinguishing features of the modern type of oil lamp are the aeration of the flame and the use of the glass chimney. It is a curious fact that both were invented long before they came into common use. Leonardo da Vinci (1452—1519), surely one of the most versatile men of all time, discovered the value of an upward draught and to him is due the idea of a metal chimney above the flame, but it was not until two hundred years or so later that Quinquet, a French apothecary, hit upon the notion of substituting a glass chimney—even then it was not put around but over the flame (Ilin, *Turning Night into Day*).

In the old open-flame lamps of the eighteenth century, surviving into the nineteenth, the size of the wick was limited owing to the fact that a large thick wick would mean incomplete combustion through the centre being inaccessible to the air. To a certain extent, this difficulty was met by the production of the flat wick, which was substituted for the old round wick in certain types of lamps early in the nineteenth century and has survived until the present day in many forms of the paraffin lamp. As early as 1784, however, Argand had got over the difficulty by the invention of the circular tubular wick inserted between the inner and outer tubes of a double burner, a current of air through the inner tube thus reaching the inner surface of the circular flame. This did not, at first, give the brilliant flame he had expected, but the placing of the broken neck of a bottle over the flame had marked results. So, as in the case of so many valuable inventions, the idea of the glass chimney was born almost by accident. It is strange that generations of more or less scientific craftsmen, with full knowledge of the value of a draught in increasing the power of a fire, took so long to hit upon the idea of applying the principle practically to artificial light.

Notwithstanding the incidence of these two valuable discoveries, the old open-flame lamp, with its small solid wick, held its own for common use well through the first third and to a large extent the first half of the nineteenth century, and it was not until the cheap paraffin lamp made its appearance that there was any marked departure from primitive principles.

The Argand lamp itself was a lamp based rather upon the principle of the bird fountain, the oil being supplied to the wick by gravity, the rate of feed being regulated by a valve, a principle which had already been propounded as a means of improving the flow of oil to the burner by one Cardan in the sixteenth century. This principle, in addition to being applied to lamps of a cheap type in tinplate, &c., is that of the often highly ornamental (though not necessarily tasteful) 'astral' lamps of the early nineteenth century. These were usually in groups of three, the centre and side pieces of a mantelpiece set, the lineal descendants of the mantel candle sets of pretentious drawing-rooms. The idea in these

'Argand' type lamps of having the reservoir at a higher level than the burner was effective but inconvenient, the most convenient position of the oil reservoir being obviously at the base of the lamp itself.

It was not until the end of the eighteenth century, when the first signs of the use of other means of illumination had already appeared, that inventors turned their attention to effecting any really marked improvements in the oil lamp. Gas lighting was already in existence then and the gas industry had become an established institution before most of the oil-lamp inventions came to light (in either sense!)—in fact, years before the modern principles of oil lighting came into general use. Curiously enough, the infant years of gas lighting actually saw an increase in the consumption of whale oil, probably, as has been suggested elsewhere, through the greater use of table lamps in place of candles.

The nineteenth century brought a flood of inventions, good, bad, or indifferent, many of which, even if more or less successful at the time, had a very short life. It may suffice to mention a few.

The Bude light. An Argand lamp in which oxygen was passed into the lamp —later adapted for coal-gas.

The Drummond light. An intense light produced by a jet of oxygen feeding a flame in contact with lime.

The Carcel lamp (1800). A device to get over the difficulties of oil supply, with heavy oils and a reservoir beneath the burner, which consisted of a double piston operated by clockwork, forcing the oil through a tube to the burner.

The Moderator lamp. Another attempt to improve the fuel supply with a base reservoir, invented by Franchot in 1836, having a spiral spring forcing oil up through a vertical tube to the burner. This had a considerable vogue in its country of origin, France, where it continued in use until fairly recently.

The Holliday lamp (1820), burning oils from tar, the oil being vaporized by the heat of the flame and burned through small orifices in a rose burner. (This is the type of the naphtha lamp used on street stalls.)

The Oleostatic lamp. There were several inventions in this group, in which, by an arrangement of chambers, the displacement of brine and oil was made to force oil to the wick by hydraulic principles.

The Diacon lamp (U.S.A. *c.* 1840). Clockwork in the base actuated a pump forcing oil through a tube from the reservoir to the wick, as in the Carcel lamp.

The Hitchcock lamp (1868). A flat wick kerosene lamp in which a fan, run by clockwork, forced air into the flame.

Lucigen, Doty, and Wells lights, in which the oils were forced from the reservoir by high pressure through a spiral heated by the flame of the lamp, the heated oil being ejected partly as vapour and partly as fluid and burning with a large and highly luminous flame. This principle, of course, survives in the modern contractors' flare lamps of the Wells type.

Far more important than all these mechanical contrivances, considered necessary to improve the consumption and efficiency of the heavy oils then in use, was the search for a better fuel, intensified and stimulated by the growing scarcity of whale oil. Lard oil, benzolene, and camphene all had their day. The most promising of the substitute fuels of the eighteen-thirties was colza oil, made from rapeseed, still a constituent of bicycle oil, &c. This was the fuel principally used in oil lamps from the time of its introduction (*c.* 1830) until the eighteen-sixties. Towards the close of the eighteenth century, the idea of burning naphtha, obtained by distilling coal at low temperatures, had been mooted.

The discovery of paraffin caused a revolution in both lamp and candle illumination. It was first discovered by Reichenbach and Dr. Christison, simultaneously, in 1830; in 1848 James Young produced it from mineral oil in Derbyshire, and its use, both as a lamp fuel and for making candles from paraffin wax, practically commences from that date.

Despite the obvious advantages of the new fuel, paraffin lamps did not at once 'catch on'. The pioneer country in their common use was Germany and most of the earliest paraffin lamps were of German origin.

In 1859 Colonel Drake discovered petroleum oil in Pennsylvania, and with the cheapening of that oil hundreds of American patents for paraffin lamps followed in the next decade. While most of these have died a natural (or unnatural) death, the net result of the efficiency of the new fuel, fed to the modern wick (circular or flat) by capillary attraction alone, was such as to effect the virtual extinction of the old crude-oil lamps. In 1865 Hinks, an Englishman, introduced the duplex burner and other improvements followed. Side by side with the development of the paraffin or kerosene lamp itself came the general recognition of the value of the glass chimney, nearly a hundred years after its first introduction, and this combination of better fuel, improved wicks, and the draught-producing glass chimney is the basis of the domestic oil lamps of to-day.

Kitson, before Welsbach had perfected his incandescent gas mantle, brought out an oil lamp having a platinum gauze mantle; subsequently the Welsbach mantle itself was substituted in the Kitson, Empire, and similar incandescent oil lamps. In these lamps, burning mineral oil, the oil is vaporized and mixed with air in a Bunsen burner before being burnt.

Thousands of years of oil lighting had passed and the 'perfect' oil lamp had only been evolved when its obsequies were already being prepared in all parts where the benefits (or otherwise) of up-to-date civilization were available. Gas lighting had been in being for more than half a century and, if difficulties of distribution confined it to urban areas, no town or large village in Britain was without its Gas Works by the time the paraffin lamp became generally available; by that time, also, experiments with electric lighting were already in progress, soon to achieve practical results and ultimately to facilitate the carrying of an entirely different form of lighting into the most remote areas. To-day, in North Africa for example, one may see the ancient float-wick lamps of the mosques

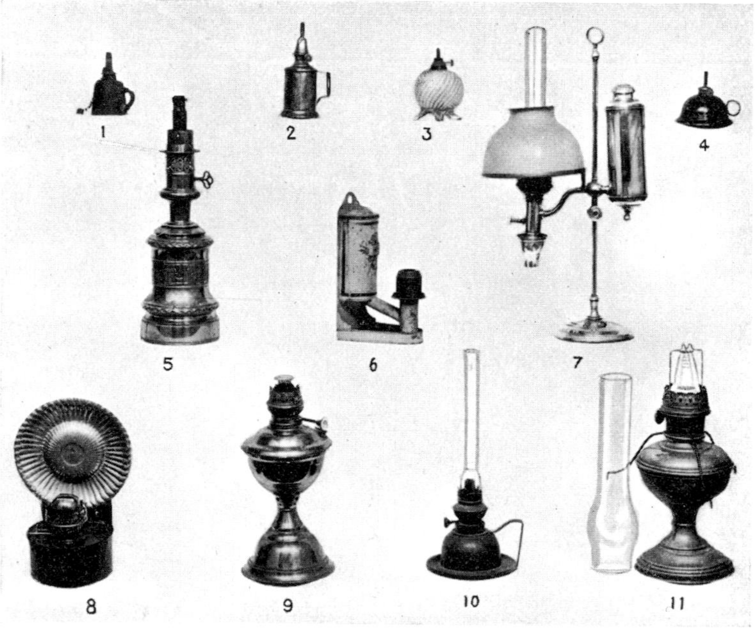

(*a*) MODERN OIL LAMPS

(*b*) SOME AFRICAN LAMPS

fitted with electric light bulbs and electric light in use far in the interior of Morocco! So passes, or is passing, the oil lamp, an age-old servant of the human race, though, like the horse, it may yet survive for generations or centuries by virtue of its comparative independence, in times of emergency or under circumstances of isolation.

[*See Plate XXIV.*]

CHAPTER XXIII

GAS LIGHTING

THE history of artificial lighting is full of unanswered queries. Why, for instance, did not the Romans, with their engineering genius, improve upon the primitive oil lamp? Why did it take until the nineteenth century to hit upon the chimneyed lamp? Why were not the mineral oils, known by seepages for an indefinite period back in Mesopotamia, used for lighting? Why, too, when the ancients knew the illuminating properties of natural gas, was not the knowledge made the basis of some use of gaseous lighting at some time or other long ago? Chinese necromancers are said to have used the gas from 'fire wells' in the province of Tsee-Leiou-Tsing by collecting it in skins, in which pin holes were made; the jets of gas issuing from these were lit (Chandler, *History of Lighting by Gas*).

Jean Tardin, a French doctor, published a book in 1618 describing experiments he had made in producing artificial gas from heating crushed coal in a closed vessel; he was led to these experiments by tracing the gas from the 'fire well' at Grenoble to bituminous coal-beds in the district. In 1667, too, Thomas Shirley gave a report to the Royal Society on 'A well and earth in Lancashire taking fire at a candle', based on some experiences he had had eight years before at Wigan. Here, in consequence of a tale that the water by a certain spring burnt like oil, he and others visited it and applied a candle to it, with the result that a large flame appeared and burnt the foot of a tree nearby. Shirley proved that the water itself extinguished the flame of the candle and that the bigger flame proceeded from gases bubbling through the water; after drawing away the latter, he got the flame from the earth.

Dr. Clayton, following similar experiments with so-called 'burning wells' in the same district at the end of the seventeenth century, followed Tardin's steps by tracing the origin of the gas to the underlying coal-beds and distilling it from the coal itself. Storing it in bladders, he pricked holes in these and applied a light, thus finding a means of illumination infinitely superior to those prevailing. Curiously enough, he made no practical use of his discovery, nor did Dr. Stephen Hales, who published accounts of similar experiments in 1726, Dr. Watson, who published an account of his observations on coal-gas in 1767, or George Dixon, who experimented in 1760.

In 1764 a Frenchman, Jars, had the idea of piping gas from a colliery in the Lyons district, but an accidental explosion put a premature end to the idea. A year later a Mr. Spedding, a Whitehaven colliery manager, not only lighted his own offices but offered to supply Whitehaven; the proposal was turned down. The early history of attempts to use gas as an illuminant is a long-drawn-out story of suspicion and apprehension delaying its introduction. A Dutch professor, Minckelers, who distilled gas from powdered coal and lighted his lecture-room with it in 1784–5, is, on the strength of this, sometimes claimed to be the discoverer of coal-gas for illuminating purposes. The first experiment of lighting a public building with gas was actually carried out by Diller, a German or Dutch scientist, in 1787.

In 1792 Murdoch, at Redruth, in Cornwall, took practical steps towards gas lighting, distilling gas from coal in iron retorts and burning it at the end of an open tube. Later, the tube was closed at the end and holes pierced laterally to act as burners. It is said that this idea was the result of an accidental discovery, as so many inventions are: using his wife's thimble to close the end of the tube and the thimble having been worn and holed, tiny gas jets resulted. In 1798 the first commercial lighting experiment took place when Boulton & Watt's foundry at Soho, Birmingham, was lit by the new illuminant. At the same place, a public display of gas illuminations was made in 1802 to celebrate the Treaty of Amiens and attracted public attention.

Meanwhile, Le Bon in France and Becher in Munich were working on gas lighting and Le Bon gave an exhibition in Paris in 1802, having taken out a patent for making an illuminating gas from wood. (Gas, incidentally, has been produced from wood in recent times, notably by Messrs. Chivers at their jam factory at Histon, Cambs., and by certain Gas Works during a coal strike a few years back.) In 1804 William Henry experimented on wood, peat, oil, wax, and different kinds of coal and with a view to the distillation of gas from tar; he published the results of his successive experiments during the following twenty years.

In 1803 Winsor, a German and a prominent figure in gas pioneering, lit the Lyceum Theatre by gas, and two years later it was used at the cotton mills of Phillip and Lee, Manchester.

Meanwhile attempts were made to produce gas from oil and distribute it in a similar manner to coal-gas. John Taylor took out a patent to this end in 1815 and between 1819 and 1830 several oil-gas companies were floated in provincial cities. An attempt to start one for London was killed by a Select Committee of the House of Commons appointed to consider the project. In 1827 the Edinburgh Oil-Gas Co. turned over to coal-gas and installations in general proved to be failures.

The earliest form of true burner was the 'rat-tail', consisting of a metal tube closed at one end, which was perforated with a single hole. By 1808 the 'cock-spur' type of burner, in which a tube end, sealed and pierced with three small holes, gave a triple flame, and the 'cockscomb', in which the holes were more numerous, were evolved. The former gave a light of about one candle-power for every cubic foot of gas burned per hour. In 1816 came the batswing burner, consisting of a small pear-shaped steel burner, about $\frac{1}{16}$ inch in diameter, having a slit at the top about $\frac{1}{40}$ inch in width, and four years later the 'fishtail' or early form of union jet was evolved—a burner in which two jets were united to form a single flat flame.

Fifteen years passed from the actual inception before the real start of public lighting came—an installation in the city of London in 1807 being followed in quick succession by the lighting of Pall Mall in 1809 and Westminster Bridge in 1813, the adoption of gas lighting in the streets by Westminster parish in 1814 and a rapid spread of the method, despite violent opposition by chimney sweeps and others, during the following years. The Chartered Company (promoted by Winsor), the forerunner of the present great London gas undertaking, the Gas Light & Coke Co., was formed in 1810; Dublin had gas lighting in 1818, Paris in 1819, and Sydney, Australia, in 1841. Judging from the dates of the formation of British gas companies, large and small, there must have been a tidal wave of formations in the United Kingdom in the 1830–40 decade. Some idea of the importance attached to gas engineering at this stage, however, may be gained from the records of the Shepton Mallet Gas Co., who, in their early years, were paying their works manager (who carried out, no doubt, a multiplicity of gas duties) the munificent sum of nine shillings a week! (Possibly, of course, the job was a part time one.) In other countries gas lighting was advancing in more or less parallel fashion. In addition to the places already mentioned, Philadelphia, U.S.A., for instance, after opposition, adopted gas lighting in 1835.

The middle of the nineteenth century saw fresh experiments in burners. Following the early adaptation of the Argand principle to gas lamps by Clegg in 1809, in 1853 Frankland devised an Argand type of burner consisting of a metal ring having a series of holes from which gas issued, fed by an air draught which passed between an inner chimney (enclosing the burners) and an outer chimney extending below the other. A base plate cut off this draught, which then passed beneath the inner chimney to the interior of the ring, having been heated in its passage. Other and later developments of Argand types of gas-burners, in which a current of air is fed to the inside of the flame, belong, however, rather to the sphere of heating than of lighting. Bowditch (1854) and Siemens (1878) brought out 'regenerative' burners which gave an increased candle-power per cubic foot of gas (7 to 10 candle-power against the 1 to 2½ of the fishtail burner).

The union jet type of burner was improved in 1880 and provided with an 'economizer' in 1890.

Just as with the oil lamp, improvements were confined more or less to details (regenerative burners, economizers, clay or porcelain and afterwards steatite tips to the burners, &c.) until a powerful incentive arose in the form of a threat to the very existence of the industry. Electric light was already on the way when, in 1881, Clamond produced an inception of the incandescent idea by hanging baskets of calcined magnesia under inverted Bunsen burners (the

(By the courtesy of the 'Bristol Evening Post' and Bristol Gas Co.)

THE DEVELOPMENT OF THE GAS BURNER

Bunsen burner flame, fed by air, being very hot but of little illuminating power). Kitson, in 1885, evolved a platinum mantle for the paraffin lamp. Finally, Dr. Welsbach, after some years of experiment from 1884, introduced his well-known incandescent gas mantle—a 'stocking' of silk or cotton fabric impregnated with rare earths (thoria, ceria, &c.)—which revolutionized gas lighting and has of recent years led to the substitution of calorific value for illuminating power. First introduced on a commercial scale in 1893, it was originally upright and is applied still in that form to modern types of the paraffin lamp such as the 'Aladdin'; in gas lighting it took an inverted form in 1905 and became shorter to conform to the lesser length of a flame ejected contrary to its natural tendency. This inversion was brought out by the desire to radiate the light downwards, in the direction most needed, instead of upwards and was the result of experiments by a number of lighting engineers from 1896 to 1903, the principal credit for the ultimate evolution of the inverted incandescent gas-burner being due to Bernt and Cervenka (Prague) and Ahrendt (Berlin).

The above is a brief summary of the story of gas lighting itself, improvements and alterations in the methods of gas production hardly being within the scope of a book of this character.

Mention must be made here of some other gaseous lights. Naphtha has already been referred to in a previous chapter, since it is really an oil fuel vaporized in the lamp itself.

Acetylene, a hydrocarbon gas generated from calcium carbide decomposed by the action of water, is said to have been discovered by Edmund Davy in 1836 or by Berthelot, who made it known in 1862. Its use as an illuminant dates from about 1892, but acetylene lighting has been confined mainly to installations in which the gas has been generated more or less on the spot, as in contractors' roadside lights, bicycle lamps, and country house installations.

Butane, in liquid form, vaporizing into gas on release, has been used abroad in places where a coal-gas supply is unobtainable and has recently been introduced under similar conditions in this country. The idea of having 'portable' gas was anticipated early in the nineteenth century, when a company was established to manufacture gas to be put into cylindrical vessels and conveyed to the houses and shops of the consumers, but the enterprise was then a failure.

[*See Plate XXV.*]

CHAPTER XXIV

THE COMING OF ELECTRICITY

THE author is not concerned to trace electrical development as such nor is it possible, in a work of this kind, to enter into the technical details of electric lighting. The story of the lamp, however, could not be told without some reference to the light which, by its universality, is doing more to cause the disappearance of primitive oil lamps than anything else; the story could not be *complete* in these days of day-to-day development.

As early as 1808–9 the possibilities of using electricity for producing light were realized, but the arc lights produced in the laboratory remained there for many years after. Save for exceptional and experimental applications, it was not until 1876–7 that the first definite electric light appeared, in the form of Jablochkoff's 'candle', consisting of two carbons placed side by side; it was used for street lighting in Paris, London, &c. In 1877 Brush invented an arc lamp, but these early types were open to the air and only partially satisfactory. For ordinary use the arc lamp was quite unsatisfactory, particularly because of its size and the constant attention needed, and it was not until the introduction of the filament lamp that electricity came into its own for general lighting purposes.

The credit for the first introduction of a practical electric lamp is the subject of keen controversy between England and the United States of America. English electricians claim that Swan, of Newcastle-on-Tyne, was first in the field—Americans, that he was anticipated by Edison. In any case, both had been preceded by experimenters such as Grove and de Moleyns (who made platinum wire lamps in 1840), Starr and King (who suggested metal or carbon filaments in a vacuum, in 1845), and Lodiguine (who made a carbon filament nitrogen filled lamp in 1873). Swan, however, had been experimenting since 1848 and in 1878 he exhibited, at Newcastle, a lamp in the form, more or less, of a cucumber, with a carbon filament in a vacuum. In 1880 Edison, who had taken out a comprehensive English patent in 1879, produced a 16-candle-power lamp with carbonized bamboo filaments. The lamp itself, with a pear-shaped bulb, closely approached the form of the modern incandescent lamp. In 1882 Edison opened the first public electric supply station, in New York city.

In 1893 came the Jandus arc lamp, a closed lamp and the first really successful lamp of the arc type.

In 1897 the Nernst (incandescent) lamp was introduced, with a filament of rare earths. Osmium filament lamps were produced by Welsbach, 1898–1902, tantalum in 1905; tungsten, first tried in 1884, became the standard material for filaments from 1909. Despite the constant improvements in incandescent lamps, however, the arc lamp (developing into the 'flame' arc in

1897) remained supreme for high power lighting until the Great War. Just before that time (1913) the gas-filled incandescent lamp arrived and the situation began to alter. To-day, incandescent filament lamps are capable of giving 2,000-candle-power illumination and are the type used for flood lighting and intensive lighting of all kinds.

Luminous 'discharge' lamps, in which an electric current discharged into a gas-filled reservoir causes a diffuse glow, had been experimented with for a long time but it was not until 1900 that the mercury vapour lamp—a high-power lamp giving a greenish light—was patented. The recognition of the value of this lamp and the sodium vapour lamp (giving a yellow light) for street and factory lighting only dates from about 1932.

Coloured light 'tubes', as used for advertising purposes, depend for their colour on the gas used (e.g. neon, red; argon, blue; carbon dioxide, white, &c.). These, however, are 'byways' rather than on the highway of a history of purely lighting appliances.

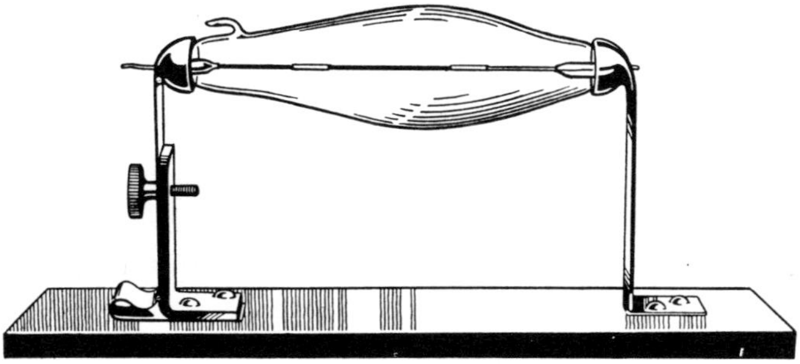

SIR JOSEPH SWAN'S FIRST PRACTICAL INCANDESCENT LAMP

Made at Newcastle-on-Tyne in 1878, and patented in 1880. The lamp consists of a rod of carbonized parchmentized thread both ends of which are slotted into platinum lead wires which are passed through the glass bulb for connexion to the current supply wires. The bulb is a vacuum, the air having been drawn off through the 'pip' at the top left-hand side.

Sketch from the actual lamp in the Ediswan Collection.

(By the courtesy of the Edison Swan Electric Co., Ltd.)

SECTION III

'SIDELIGHTS'

CHAPTER XXV

ANIMAL LAMPS

At first thought no way of lighting a home, however primitive, seems possible without a fire. So far as can be ascertained no means of illumination actually preceded the discovery of the ways of producing fire, but there is no reason why, in some parts of the world, night life could not have flourished—or at any rate existed—without a flame at all!

An early account of tropical America describes a great beetle 'rather smaller than a sparrow' whose 'two stars close by its eyes and two more under its wings' furnished sufficient light for the natives of the area to spin, weave, write, and paint by it. The Spanish invaders did not disdain to use this primitive lighting appliance on their hunting expeditions, the beetles, attached to their thumbs and big toes, not only providing light for the hunter but attracting the hunted as well. This may be the same kind of beetle as that sent from the Isthmus of Tehuanapec to the Smithsonian Institution at Washington, which was said to have emitted a sufficiently powerful light to enable a person to distinguish objects in dark rooms and read small print at six inches distance (in which case the beetle was presumably held in proximity to the reading matter, as a sort of insect reading lamp). A few beetles are supposed to have been sufficient to light a room and the night traveller wearing one on each foot could see his way—and snakes. The sale price demanded by Indians who caught these handy little creatures was a few cents per dozen, which seems to make them an economical, if low-powered, form of lighting.

These same insects were used, just as more artificial forms of lighting are, for festive illuminations. The native women, dancing the fandango, wore them fastened in their hair or head-dresses or made garlands of them, the bluish-white light giving a fantastic beauty to the gay dresses and jewellery of the dancers.

We in Britain, of course, have our beetle illuminant too, but, in the case of the glow-worm, the egg, larva, and pupa are all luminous. Feminists may dwell with delight on the fact that it is the female who is the most 'shining light'—but analogies are sometimes misleading! In any case, we do not seem to have found a practical use for the light of the glow-worm at any period of our history.

The firefly, a better-known living lamp, is less individual in its habits or application, though a group of New World species (known as Pyrophori) are 'three candle-power' insects (with two luminous patches on the chest and one on the abdomen) giving a pleasing and comparatively strong light in which Dame Nature has forestalled the coloured lights of the electrician—the chest lights being greenish and the lower one orange. Generally, however, when fireflies are put to use for their light, they are not employed as single units, the light of

European fireflies being only that of large sparks. A mass of fireflies, such as is often seen in Italy, is a sight to be remembered—a mass of brilliant living sparks through which one walks without apparently disturbing them.

In the West Indies fireflies have been used as lighting appliances in at least two forms of lanterns—one an affair of three diminishing stories made by setting small rods in square pieces of wood with truncated corners (having a door to each story), the other a perforated tree gourd. In Java a wooden dish device, with wax in the bottom to which fireflies were stuck, having a lid pivoting on one end, is said to be a 'dark lantern invented by a burglar'! A reserve supply of insect 'bulbs' for this portable light was carried in a cane tube.

Farther east, fireflies have functioned as illuminations at Japanese social functions—whether in tea-gardens or in private houses. Sometimes they became the illuminating power of a 'cage' lantern; at others they were released in large numbers to form a scintillating cloud of sparks as in their natural state. Firefly-catching in Japan, like beetle-catching by the Central American Indians, was a business. Girls pursued them with fans and boys caught them with wands having wisps of yarn attached, while singing old folk-songs the while (though it is not clear whether these were for the amusement of the catchers or the fireflies). Alternatively, men were employed to catch them by striking branches with poles or gathering the fallen insects into mosquito bags.

Other denizens of the animal world have figured as lamps or candles, but in such cases they provided the lamp or the fuel rather than the actual light itself. The oily candle fish of British Columbia, for instance, became an Indian candle by having a strip of bark passed through it as a wick, and the stormy petrel was used (in the Orkneys and Shetlands, for instance) in a similar way by pushing a wick down the dead bird's throat. In Venezuela, the nestlings of a nocturnal bird, the guacharo, are boiled down into fat and oil to be used for cooking and lighting.

The bodies of sea-urchins, filled with fish oil and provided with wicks, have been used as lamps in Japan, and the skull of a sheep forms one of a group of primitive lamps in the Smithsonian Institution collection. These, like the use of the shells of sea animals as lamps, are simply cases of utilizing natural bodies as receptacles for oil. Fireflies and luminous beetles, on the other hand, are Nature's living lights.

CHAPTER XXVI

LANTERNS

It is impossible to fit the lantern satisfactorily into the framework of the history of the candle or that of the lamp, since it belongs to both and has no definite place in either. Actually, of course, a lantern is not so much a lighting implement in itself as a cover or protection for one and containing one, and that contained lighting implement may be a candle, a night-light, an oil lamp, or, in modern times, a gas or electric appliance. It (the lantern) may vary in construction from an elaborate and ornamental fitting in which the actual light is almost an excuse around which to mount a work of art to a hollowed-out turnip with a candle inside (such as one may still see in the hands of the children of rural England sometimes).

Primarily, the purpose of a lantern is to shield a light from the effects of wind and weather and thus enable it to be used outdoors. This definition covers most forms of lantern types, with the exception of a few indoor protected lights which it is difficult to place elsewhere. It necessarily includes, as lanterns, ship, carriage, and railway lamps, street lamps, and all such outdoor devices for the protection of an enclosed light.

The first true lantern, like the candle, appears in Roman times, so far as Western civilization is concerned at any rate. The Roman lantern, like the rural lantern of recent days, was a horn-windowed affair. One found at Herculaneum was cylindrical, with a hemispherical top, and made of sheet copper, except the two main supports, which were cast. The bottom was a flat circular copper plate supported by three balls and turned up all round the edge. The top and bottom were connected by rectangular supports and also by interior uprights. The top cover was removable and had air holes; the illuminating medium itself was a small oil lamp. Both hand and hanging lanterns were used by the Romans and were found at Pompeii and Herculaneum.

A primitive variation of Roman times is to be seen in the museum at Alexandria, in the form of clay lanterns, remarkably like certain night-light containers of modern times, with an open front and intended to carry a small pottery lamp inside; these lanterns are cylindrical or square, plain or decorated, and mostly adapted for suspension.[1] Something similar is to be seen in a solitary Romano-British example in the British Museum, which, however, has perforated sides which seem to foreshadow the perforated metal lanterns of more recent times.

Once again, in the use of the oil lamp inside these lanterns, we have the spectacle of ancient Rome anticipating modern times. The oil lamp is not the light inside the lantern of the Middle Ages or in the periods immediately following,

[1] Cf. Maltese pottery lantern in frontispiece. These lanterns are said to have been used at shrines, out of doors.

save perhaps in southern Europe, where Roman tradition persisted, and in the East. Even the great lantern of Eddystone lighthouse was furnished with a multiplicity of candles, as any visitor to Plymouth Hoe may see. Candles were the almost universal lights of the lanterns of the West for many centuries.

Chronology, or even an arrangement in sequence, is difficult in the case of lanterns. The idea of protecting a light in the open is a natural one which would be forced upon the user by circumstances, and it is therefore not merely unnecessary but actually illogical to suppose any sort of common origin or even such a limited number of separate origins as one has to concede for the lamp. For this reason, forms and materials run side by side and precede or succeed each other with total disregard for the historian. Naturally enough, the materials from which the framework of a lantern is made vary with the supplies available for use economically and efficiently in various parts of the world and include copper, bronze, iron, tin, steel, brass, leather, wood, bamboo, and clay.

The principles on which lanterns are constructed fall mainly under four heads:

1. Simple perforations of the casing—notably in the sheet iron lamps and gauze lanterns of Europe and the openwork brass lamps of the Near and Middle East. In this category too may be included the clay lanterns of some parts of Africa.
2. Lanterns in which the skeleton frames are fitted with plates of thin horn —a European device.
3. Lanterns in which the actual covering is of paper or textile material—confined, with few exceptions, to the East and including the collapsible silk, linen, and paper lanterns of Persia, Egypt, and China.
4. Lanterns fitted with glass sides or fronts—widespread and all more or less modern. These may be grouped into (*a*) lanterns with a glass front only, or with a glass front and small insertions of glass at the sides, and (*b*) lanterns having a large area of glass.

The open-sided bronze, stone, and pottery lanterns of Japan might be assigned to a fifth category, but they closely resemble some of the textile covered types and have themselves, on occasion, paper or textile coverings fitted to their open spaces. The Japanese stone lanterns, incidentally, present some sort of parallel to the medieval 'Lanterns of the Dead'—tall stone towers placed, like lighthouses, in French medieval cemeteries (particularly in Aquitaine) and containing a light in the openwork top story, raised and lowered by a cord. These, according to tradition, were lit at the moment of death, as an alternative to the passing bell, and remained illumined until after the funeral.

1. Obviously the perforated metal lanterns are of limited utility, a large proportion of the illuminating power of the contained light being obscured and wasted by the area of metal intercepting the rays. In the sheet iron and bronze specimens of Europe (which are candle lanterns) the principal and almost only merit is durability. They are commoner on the Continent than

(*a*)

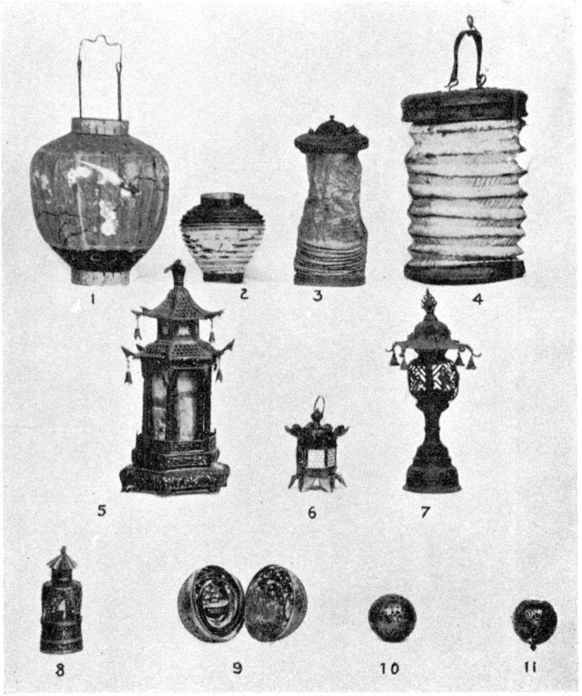

(*b*)

LANTERNS

in England (particularly in Scandinavia, Russia, and central Europe), but the principle of the perforated surround is present in the 'nightguards' in which rushlights were placed, in English bedrooms of the eighteenth and nineteenth centuries, these differing from the lanterns in that they had no tops.

In this category must be placed the gauze lantern of fairly recent days, particularly useful in barns and places where chaff or other inflammable dust was floating about. The thin mesh of a gauze lantern was of course less obstructive to the light than the sheet-iron sides of the perforated lantern proper.

Lanterns outside Italian palaces of the Renaissance period (as one at the Ca' Foscari, Venice, a very notable example) are of openwork metal and form a half-way house between the north and central European simple and even crude types and the ornate perforated lanterns of the East. One in the National Museum, Florence, is in the form of a cornucopia, and others of more usual type on sixteenth-century Florentine palaces show the influence of Gothic architecture. Miniature examples of these are often to be found in Italian (especially Florentine) cemeteries.

Perforated or openwork metal lanterns (or lamps as they are more usually described) of the Near and Middle East are generally of brass or bronze and often cylindrical in form. Some, usually pyramidal in form, which may be described as more or less lantern types, are more properly 'chandeliers' in that the little float-wick lamps are carried outside the framework; a number of good examples of this kind are to be seen in the Arab Museum at Cairo.

Among Persian hanging examples is to be found a complete sphere, with a candle socket in the base and opening at the top. An interesting relation of this type, not for hanging,[1] is the 'Rolling Lamp', in which there is no opening at the top or candle socket, but the interior is occupied by a gimbal arrangement of concentric metal rings in the centre of which a small central burner enclosed oil lamp is so suspended that it remains upright, in whatever position the sphere may be (Plate XXVI, *b*, Nos. 9 and 10).

2. *Horn Lanterns*. As already stated, the idea of a lantern with horn 'windows' originated in ancient Roman days and whatever lanterns may have been used in the Dark and early Middle Ages would certainly have carried on the idea, since the use of horn was prevalent among northern races from a very early period. Right up to the middle of the nineteenth century A.D. horn, because of its lesser degree of fragility compared with glass, held its own in rustic lanterns and was used in frameworks of iron, wood, or leather (but mostly the first-named). At first sight the word 'lanthorn', applied in olden times to the lantern with horn windows, suggests that the horn had given the name, but medieval references to lanterns (e.g. in a poem of the thirteenth century, 'Cursor Mundi', and in Chaucer's 'Legend of Good Women') are in a form which suggests the modern 'lantern' rather than 'lanthorn', and the latter would therefore appear to have been a rustic corruption arising from an association of ideas. (As a

[1] But the author has recently seen one adapted for hanging and unsuitable for any other purpose.

matter of fact, both the Saxon lantern in Strutt's *Chronicles of England* and the fourteenth-century lantern on f. 91 of the Luttrell Psalter suggest rather the perforated or openwork type, which, being of northern origin, might well have been brought in by the Saxons.)

Alternatives to horn, before the general adoption of glass, were mica (which came into England in the sixteenth century under the name of 'Muscovy glass'), bladder, and oiled paper. An interesting example of a small travelling lantern is a mid-nineteenth-century specimen, folding up into a tin box, the top and bottom of which, with two hinged mica 'windows', form the four sides of the lantern, with a hinged base and top and a small recess at the back of the box for reserve candles. Folding lanterns with mica windows, on a somewhat larger scale, were served out to the United States forces in the Great War.

3. The place taken by thin plates of horn in the Western world was taken by paper in the Far East and textile materials in the nearer Orient. It has been suggested that the collapsible lantern originated in Persia, where a concertina form having a bronze or brass top and bottom and a linen folding body (usually reinforced by wire hoops) on the same lines as the well-known Chinese paper lantern was in use. This form has also been used in Turkey and Egypt in recent days (more modern examples being with brass top and base) and was particularly handy for the traveller who, when camping, suspended it from a support of three staves (*vide* Burden, *Oriental Customs*, &c.): a candle was the illuminant.

It is in the Far East, however, where this type of lantern has been and is most prevalent, especially in China, though the materials used are silk and paper. Silk and paper fillings or screens are not there confined to the collapsible lanterns. They appear also in non-folding lanterns on frameworks of metal and wood in China and Japan, especially the latter. Chinese lanterns are usually for candles; the commoner collapsible specimens are of paper, the better of silk on a bamboo frame, the bamboo ribs being attached at the ends to wood. In Japan collapsible paper or silk lanterns with prickets for the candles take different forms or have different attachments for various purposes—as the *bajo* (for riding) and the *umihari* (for walking), handled lanterns, but there are also non-collapsible wood frame lanterns, paper-covered, for household use, with saucer lamps or candles inside.

Non-collapsible lanterns with painted paper sides are referred to in *Oriental Customs* (Burden) as being used in modern Egypt, but the paper in this case seems to have been in substitution, in certain cases, for the more usual *reed* sides in such lanterns.

Lanterns with oiled paper sides were occasionally used in Europe as an alternative to horn lanterns; the Guilles Allès Museum in Guernsey has a linen tent-like lantern native to the island.

4. As with Indian temple lamps, there is no end to the varieties of lanterns having glass 'windows' or sides. They fall, broadly, into two categories—those

in which the metal work is a mere supporting frame, and those in which the glass window occupies merely the front (or perhaps the front and sides) of the lantern.

The former group includes the lanterns which follow the lines of the old lanthorn, with the substitution of glass for horn, circular lanterns of many kinds, ships' lanterns (other than navigating lights), the ordinary street lamps, and last, but not least, the hurricane lamp, known to us of the Western world principally for its use in road and other contracting work but a modern domestic light in the native dwellings of India and Africa.

The second group (to which many of the early hand lanterns belong) includes navigating lights, dark lanterns of various kinds (including the old policeman's bull's-eye lantern), carriage lamps, railway lamps of various kinds, wall-lanterns, the old postman's lantern, &c. Generally they contain oil lamps.

Special lanterns were sometimes associated with ceremonial use. The city of Chichester, for instance, has a municipal lantern for processional purposes, in the form of a large globe mounted on a staff. Similarly, in the Hanseatic Museum at Bergen (Norway), there are flat clock-faced lanterns for candles, to be mounted on poles and carried before funeral processions of the Hansa merchants.

A curious instance of snobbery is evidenced by outside hanging lanterns of old Guernsey. Only members of the 'sixties' (the 'upper ten' of the island) were permitted to use 3-candle lanterns! (The next stratum in Guernsey society being the 'forties', presumably they would be allowed two candles?)

Night-light lanterns. Night-light containers are only lanterns in the form of their construction, but in that respect they fall into that category, more or less. Prototypes of modern 'night-light' arrangements are the clay pots with open fronts already referred to, several specimens of which are in the Alexandria Museum, and a specimen of which, probably of later date, the author obtained from Malta; these held small clay lamps and are strongly suggestive of a modern type of night-light container which takes the form of a small pottery 'hut' with an open front. In this country, however, the more usual form, where the night-light was put in any covering at all, was usually a glass dish, mounted on a pedestal, on which the night-light was set beneath a glass cover, plain or coloured. In Germany and Switzerland the night-light covers approach more nearly to the lantern form; one type is a square iron frame with transparent porcelain panels carrying etched or embossed pictures of typically sentimental German domestic scenes while another is more elaborate and pleasing, being a three-piece arrangement consisting of a porcelain cylinder with scenic or floral decorations in colour and gold, set on a circular shallow dish and surmounted by a small 'teapot' for warming 'sweet water, in the night' (according to a Swiss lady informant).

The bucket-like coloured glass fairy lamp used so much in illuminations until the advent of the coloured electric light bulb is, of course, a form of 'night-light lantern'.

[*See Plates XXVI, XXVII, and frontispiece.*]

CHAPTER XXVII

MINE LAMPS

THE mines of ancient days used lamps which were the ordinary domestic appliances or adaptations of them. As the mining was metallurgical the ancients were not, of course, up against the problem of explosive coal-gas dangers, and when there was any special adaptation of the lamp it took the form of a handy method of fixing it for use, such as a spike for handle, which could be stuck into the wall of the mine. According to Neuburger (*The Technical Arts of the Ancients*), in many cases mines were lighted by pieces of wood soaked in resin or fat and attached to the walls by means of lumps of clay (a curious analogy to the method Cornish miners used quite recently to attach tallow candles to their leather caps). At Villefranche, he records, spoon-shaped miners' lamps of lead were found in Roman mines, the hollow of the spoon being filled with oil and the straight stem used for holding or insertion in the wall. In the same mines were found ordinary Roman lamps of earthenware.

A development of the ancient idea in a primitive lamp of more or less modern date is seen in an open clay lamp, strongly reminiscent of the early Phoenician lamps, from the Sardinian tin-mines; this had a crude flat handle or projection which could be thrust into a cranny (No. 3, Plate XXVII, *b*).

Both open and closed lamps were used in various parts at various periods as miners' portable lights.[1] A rather favourite form, especially in Britain, was a miniature 'coffee pot' type, with a projecting spout, a hinged lid and a hook to enable it to be hung on the miner's cap. This was generally used with tallow and was in vogue until quite recently in gas-free coal-mines such as those of the Radstock coal-field and Fifeshire.

A very curious lamp, of a flat spheroid type with an iron loop over and a rod and hook for suspension, seems to belong primarily to Germanic lands, but is also found in France and the author has one in his collection which was said to have been made by a local smith for use in the Botallack Tin-mine, Cornwall; if this is correct, it may have something to do with the advent of some German mining engineers some hundred years or more ago. An interesting feature of this closed lamp is that the filling hole in the centre of the dome-shaped top is invariably surmounted by a small figure of a cock, standing on the small sliding cover. The flame in these closed reservoir lamps is open and unprotected.

In many gasless mines tallow dips were the general means of illumination, the dip or candle being carried on the cap, either, as in the case of the Cornish tin-mines, stuck on with a lump of clay or (as in the Radstock area and elsewhere) fastened on with a special clip. In America a spike-ended candlestick was stuck

[1] No. 13, Plate XX, *a*, is usually described as a mine lamp.

(*a*) 'TRAVEL' LIGHTS

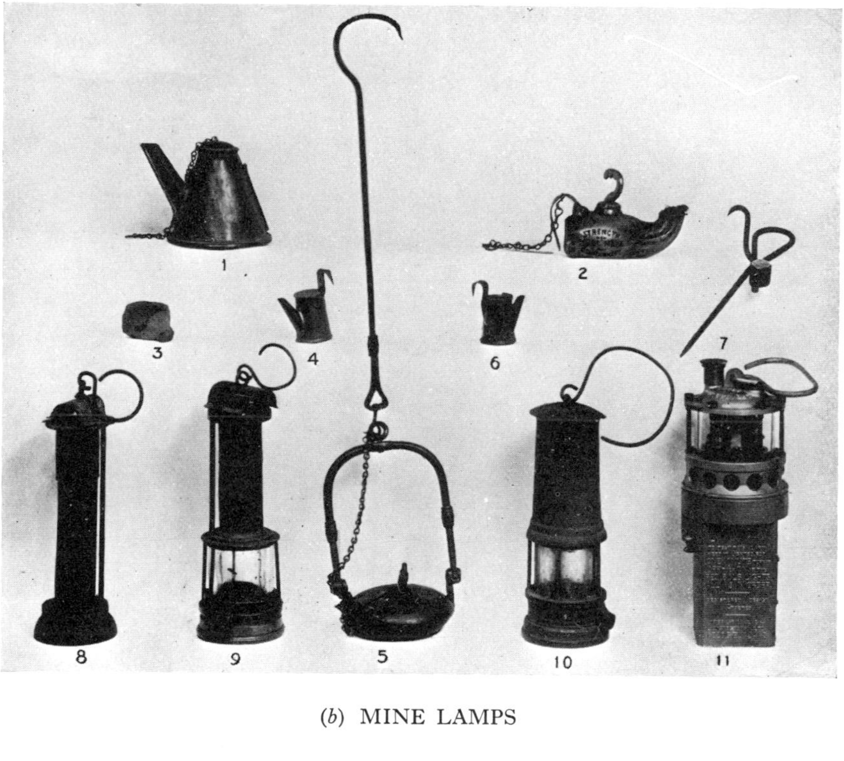

(*b*) MINE LAMPS

in the mine walls in the same way as the stem of the ancient lamp to which reference has been made above. Dips are still used in some of the Cornish tin-mines but are now generally made of paraffin wax; the wick remains of the old type, unplaited, as it gives a larger flame, and is less liable to be put out by a draught or dripping water than the modern plaited wick. Cap lamps, acetylene or electric, are gradually replacing them, however.

In the collieries and other mines of Saxony, an open-fronted wooden lantern for an oil lamp or candle has been used in fairly recent times (and may still be in use). Something similar has been used in certain collieries in the Newcastle-on-Tyne area. Hurricane lamps are used below ground in pump-rooms and the like in non-gaseous mines. In boiler-houses above ground, and no doubt in some places below ground as well, in the Cornish tin-mines, brown earthenware teapots with the spouts broken short have been used as quite effective 'spout lamps'.

The specialized mine lamp belongs to the history of coal-mining and the necessities of the gaseous mine. The problem of devising a form of lamp which could be used with a minimum of danger in coal-mines containing explosive gases was first tackled early in the eighteenth century by such primitive means as natural phosphorescent materials. About 1760 the 'spark mill' was introduced; this was a steel wheel arrangement providing a stream of sparks by friction. Its safety was only a comparative one and it required a man to work it continuously throughout the day wherever work was being carried out, so that its expense was considerable.

About 1813 Dr. Clanny designed a lamp consisting of an air-tight chamber into which air for combustion was pumped by bellows, water seals being provided for the incoming air from the bellows and the outgoing products of combustion, so that all air entering and leaving the lamp passed through water, so cooling the gases down to prevent the danger of explosion. While safe, this arrangement was awkward and cumbersome, again requiring an attendant to work the bellows, and it never came into general use.

Already gauze lanterns had been used in barns and such places where inflammable dust floated in the atmosphere, and, in 1815, Sir Humphry Davy, investigating the question of the miner's safety-lamp, made use of the principle of the cooling effect of metal gauze on a flame attempting to pass through it. The explosive gases in a mine ignite at certain temperatures and below these the flame is extinguished, so that, with inflammable gases on one side of a gauze 'screen' and a flame on the other, the flame will not pass through the gauze so long as it is kept below a certain temperature. Davy, after experimenting with glass and metal tubes, constructed a lamp in which the air entered through tubes of small diameter and wire gauze-covered holes below the flame; he then discovered that a wire gauze cylinder around the flame and closed at the top rendered the flame safe in a gaseous atmosphere, the mesh of the gauze necessary being finer in proportion to the inflammability of the gas. After an initial adoption of a gauze of 6,400 apertures per square inch, a mesh of 784 was eventually adopted and has been standardized since. Contemporary experimenters came to the same conclusions as Davy, whose lamp became the basis of subsequent improvements; one of these, introduced by Davy himself, was to protect the flame from draughts by shields in front of the gauze, the safety of the Davy lamp being affected by explosive air currents.

Up to 1839 these safety-lamps were gauze-screened from top to base, but, at that date, a vital improvement in the illuminating power of the lamp was effected by Dr. Clanny's introduction of glass, in place of the gauze, in the lower part of the lamp. The original 'Clanny' type was subsequently further improved by putting a 'bonnet' around the gauze upper half to render it safer in explosive air currents of high velocity. Other inventions, based on this type of lamp, include alterations in the method of air-supply and a chimney or chimneys to counteract the tendency to smoke. Marsaut, a Frenchman, in 1882 devised a lamp like the 'Clanny' type, but having two conical gauze caps, the inside one being more sharply coned than the outer; outside these gauze caps is a bonnet closing them in. The air-supply entered by holes in the bottom just above the glass of the lamp and the products of combustion left through holes in the bonnet close to its top and just above the top of the outer gauze cover. This is the basis of the oil-burning safety-lamp of to-day.

The oils used up to 1900 were generally vegetable oils and to some extent seal oil; modern lamps of the flame type use mainly mineral oil or mixtures of vegetable and mineral oils.

Whilst, owing to its usefulness for detecting the presence of gas, the flame lamp is still the standard for examiners, firemen, and deputies, electric lamps, first used underground about 1883, are rapidly replacing it for general mine use. Their great drawback has been the fact that, gas having no effect on an electric light, the lamp ceases to provide the collier with any warning of the presence of danger. This disadvantage is met, in the case of electric 'alarm' lamps, by the provision of two bulbs, one clear, the other red, the red being the alarm light. In the 'Ringrose' lamp, for example, this alarm is actuated when the inflammable gas (fire damp) surrounding the alarm is burned on a hot filament, a reduction of pressure being then caused inside a porous pot, which has the effect of collapsing a diaphragm and bringing two platinum contacts together. A sewer lamp of the same make is adapted to give warnings for asphyxiating gas and sulphuretted hydrogen as well.

[*See Plate XXVII.*]

CHAPTER XXVIII

LIGHTHOUSES

UNTIL comparatively recent times, the story of the seafarers' warning light was that of the fire rather than the lamp, and there was little difference between the beacon that warned the mariner of danger from natural elements and the beacon that warned of a human enemy. In fact, a far nearer parallel to the lighthouses of to-day than the shore beacons of the Early and Middle Ages was the 'Lantern of the Dead' in a medieval French cemetery. These *Lanternes des Morts* (traditionally used as variants to the passing bell, lit at the moment of death and remaining illuminated until after the funeral), especially built in the Charente and Vienne areas of France, were stone towers, hollow inside to permit the working of a pulley or the insertion of a staircase, and provided with openings at the top through which the rays of a lamp shone. The lighthouse analogy is somewhat marked, though there seems to be no traceable connexion between these *Lanternes* and the modern lighthouse.

Turning, however, to beacon lights, of more or less permanent institution, used to assist navigation, there are some evidences that these were in use, several centuries before the Christian era, by the Greeks and Phoenicians, if not by others, the light generally being provided by a wood fire. Homer alludes to the lighting of beacons on the shore to guide sailors in dangerous places, and Lesches (660 B.C.) mentions what appears to have been a regularly maintained lighthouse at Sigeum, in the Troad. It has been suggested that the Colossus erected 278 B.C. at Rhodes may have held beacons in its hands.

We are on firmer ground with the well-known Pharos of Alexandria (270 B.C.), in which the light was provided by beacon fires; according to Josephus, the light of the fires was visible forty miles away. This celebrated structure was destroyed by earthquake early in the thirteenth century.

The Romans erected a number of lighthouses and adopted the name of 'pharos' for them. Notable examples are the pharos at Ostia, built by Claudius, A.D. 30, and others at Puzzuoli, Messina, Ravenna, &c. The best example of which there are any traces in this country is the pharos at Dover, now within the confines of Dover Castle; a sister building, across the Channel at Boulogne, was lit, until the sixteenth century, by a lantern in its roof. Other British examples were at Folkestone, Flamborough, St. Andrews, and elsewhere on the east and west coasts of England and Wales; a foundation structure at Richborough is sometimes thought to be the base of one.

The Tour de Corduan, off the river Garonde, built at the beginning of the ninth century, was probably the first post-Roman lighthouse and this (rebuilt in the fourteenth century and again at the end of the sixteenth) was lit by wood fires until the eighteenth century, when coal was substituted; incidentally and later, it was reputed to have been the first to have a revolving light.

Where the secular authorities provided any lights at all on the coasts in the Middle Ages they were generally wood fires at first, but later (from the fourteenth century generally in this land) high standards with 'pitch pots'—the typical medieval 'standing cressets'—were adopted. There were difficulties with these, on account of their liability to extinction by wind or rain and the fact that a strong wind from the sea would carry the flame in the opposite direction to that in which it best served its purpose. Lights maintained by the medieval church (or its 'camp-followers', the anchorites) were in a different category. One of the three great works in which the monastic orders were enjoined to engaged was the assistance of travellers, and the form this took along our coasts was the provision of small chapels, such as that at Ilfracombe and the one on St. Aldhelm's Head, Dorset, from which an inside light was shown for the guidance of seafarers. Alternatively, such lights were attached to hermitages or 'anchorages', where they were attended by solitary hermits ('eremites') or anchorites, who made this their life's 'job' of Christian service. Where there was a chapel attached to the light it was generally, as at Lantern Hill, Ilfracombe (where the building still stands), and at Pendeen, Cornwall, dedicated to St. Nicholas (Santa Claus), the patron saint of seafarers as well as of children. This was not, however, invariably the case and the modern lighthouse at St. Catherine's Point, Isle of Wight, was preceded by a fourteenth-century tower used as a beacon. St. Michael's 'chair' at the corner of the tower on St. Michael's Mount is the remains of a stone lantern from which a light was displayed for the benefit of fishermen, the monks of Holy Island maintained a light, and others under ecclesiastical auspices included beacons at 'Boston stump' (the tower of St. Botolph's, Boston), Tynemouth Priory, Arundel Church, and a light at the south-west corner of Rye Church (supplemented by one on the south-east corner of the Ypres tower, as twin guides to the harbour). Where the lights were monastically maintained, as many were, the dissolution of the monasteries by Henry VIII in 1536–9 created a gap, as in the case of the care of the poor, which was not repaired for a long time afterwards. Lights on church towers, like those of early secular beacons, were usually open wood fires, but oil lamps were used where the light was shown from inside the building. The wood fire was general, with exceptions, up to the seventeenth century and survived into the eighteenth.

Such attempts as were made to provide secular lights in the later Middle Ages and after the dissolution of the monasteries were either by private owners or by local guilds, and this system obtained until private owners were finally bought out and coast lighting became a national affair. Naturally, since the main idea of the private owner was to secure a profit on dues from the shipowners, the efficiency of the light was not always the first consideration, to say the least. Wood fires came into their own again with the establishment of light 'houses' or towers; the first regular lighthouses are said to have been two at Caister, erected in 1600, but they were soon followed by one at Lowestoft in 1609 and another at Dungeness in 1618. Meanwhile, 'Trinity House' was given power to erect lighthouses in 1617, but the power was not a monopoly.

The story of the maintenance of the lights under such conditions was similar to that of the bridges. However commendable may have been the intentions of the first builders, later owners only too often collected the dues and forgot to properly repair and maintain the lights.

It seems hard to believe that the poor light of a candle had ever to suffice for a lighthouse. Caister lighthouse, established in 1600, had to depend on a single candle, or at most two, of six to the pound. The two lights placed at the North Foreland in 1505 were also candle-lit for nearly two centuries. The Eddystone lighthouse of 1698 was the first to adopt candles on a large scale, and Smeaton, in the third lighthouse at this spot (1759), continued this means of illumination, the candles being set on a corona of large diameter which is still to be seen in the tower preserved on the Hoe at Plymouth. The candles required snuffing every half-hour. Incidentally, Smeaton recorded that it was a matter of complaint that the keepers, who had to find their own provisions, were at times reduced to eating the candles! Candles were in use on the Eddystone up to 1810.

During the seventeenth century coal fires became the usual method of providing coastal lights. The lighthouse on the Isle of May, in the Firth of Forth, was lit by an open coal fire on the top of a tower in 1635 and retained it for 181 years. At the North Foreland an open coal fire replaced the candles in 1694, but the latter returned a few years later, the light of the fire being badly affected by wind and rain. St. Bees (Cumberland), in 1718, and Spurn Head (early eighteenth century) were among many others lit by naked coal fires, as well as the first lighthouse on the dreaded Casquets.

In an endeavour to economize on the use of coal fuel, wastefully consumed under the 'forced draught' of high winds, and also to improve upon the great defects of a coal fire—that of being liable to extinction by rain or wind and of virtual invisibility from the windward side when a high wind was blowing the flames away—the idea of enclosing the fire in a glass 'lantern' was adopted. An early example of such protection, if not the earliest, was at the lighthouse on St. Agnes, Scilly Isles (1681), the fire-grate of which is still preserved on Tresco; another early example was the lighthouse at Portland Bill (1716). It was a case of 'out of the frying-pan into the fire'. The glass screen was readily obscured by the smoke of the coal fire and many were the complaints as to the light not being seen. Also, in some cases, the fire, bereft of the draught, had to be blown into flame with bellows. The introduction of a form of chimney improved things a little, but there were still constant complaints. Nevertheless, it was not until nearly the middle of the nineteenth century that the last coal fire was discontinued.

Although oil had been used in some cases by the ecclesiastics who provided coastal lights in the Middle Ages, its use died out with the Reformation apparently. The resuscitation of the use of oil for lighthouses commenced in America. Boston's first light (1716) had oil as its lighting medium, and oil was adopted generally in American coastal lighting from its early stages. In 1789 there were sixteen lamps so lit in the United States, and in 1811 Argand lamps and reflectors with a revolving mechanism heralded the day of flashing lights. (Ordinary reflectors had been used as early as the seventeenth century in England.) Meanwhile, in England, Liverpool had adopted oil lamps with flat wicks in 1736, and Argand lamps came into wide use at the close of the eighteenth century. The oil used was generally sperm oil, but coco-nut oil, olive oil, and lard oil were also used, the last named particularly in the United States of America. Colza oil was adopted some time after 1830, largely owing to the growing scarcity of sperm oil; in 1846 colza oil was a half to a third of the cost of sperm oil per gallon. Oil replaced candles at the North Foreland in 1787 and coal fires at St. Agnes, the Casquets, and the South Foreland before the close of the eighteenth century. It did not replace coal on the Isle of May until 1816 or at Dungeness until 1831.

Although mineral oil had come into general use some years before 1868, it was not until that date that a suitable lamp was invented to enable it to be used in lighthouses. The principal oil-lamp types thereafter were:

(*a*) The fountain type, in which the oil flows by gravitation from the reservoir at a slightly higher level than the burner.

(*b*) The pump type, in which the oil is forced into the burners by pumps actuated by clockwork.

(*c*) Oil under pressure forced up from cylinders by a weighted piston.

In the 'Bude' lamp intense combustion was induced by the introduction of oxygen into the centre of the oil wick, but the constant care it required militated against its success.

Oil lighting spelt a great advance over the old beacon method and required less attention, some of the petroleum lamps in France running for three months unattended.

Coal gas, suitable, of course, only in certain land positions, had been used on the Continent in 1823, but was first adopted in Britain at Troon Pier in 1837. It was largely introduced, where available, by the middle of the nineteenth century and is still used at a number of lighthouses situated in towns, the advent of the incandescent mantle having, of course, greatly improved its efficiency, a light up to several hundred thousand candle-power being obtainable by its means before the close of the nineteenth century.

Acetylene has been fairly extensively used, or in some cases is used, as a 'stand-by', particularly at electrically lit stations.

As early as 1858 an experiment was made with electric lighting and an installation was made at Dungeness in 1862; it was, however, abandoned in 1874. A permanent electric installation was made at South Foreland in 1872 and another at the Lizard in 1878 (both replacing oil), but it was not until about 1875 that electric arc lighting was regarded as really practicable for coastal lighting. In 1883 Trinity House decided that oil and gas were more suitable than arc lamps except where a great range of beam was required.

To-day electrically lit lighthouses are in the majority (generally with an alternative as stand-by in case of breakdown) but by no means universal.

CHAPTER XXIX

STREET LIGHTING

In the story of street lighting we are once again faced by the ever-recurring fact that ancient Roman civilization was far more advanced than that of so recent a period as the eighteenth century of the Christian era. Even in such an outlying Roman city as Antioch, in Syria, in the fourth century A.D., for instance, according to Libanius, there was organized street lighting, the lamps being suspended on ropes—an early example of central suspension lighting!

There was little or nothing of effective public lighting in medieval Europe. In 1416 Mayor Sir Henry Barton ordered lanterns or other lights to be hung out in the city of London in winter (from All Hallows to Candlemas), a limitation paralleled by the practice, even recently, of some small communities, of taking down their street lamps in the summer and even leaving them unlit on moonlight nights in the winter. This London ordinance continued, with constant evasions and little real effectiveness, until the time of Queen Anne. Cressets (beacons or fire-baskets) were used for stationary lighting (on poles, with rude steps attached, as shown in the coat of arms of Bishop Beckington, at Wells, Somerset) and by the watch in the sixteenth century, pitch and suchlike substances being the usual fuel. Stow records that 700 cressets were used by the 'marching watch' and that every one of the 240 constables had his cresset. Each cresset had a man to bear it and another to serve it, making well over a thousand in all, apart from the 'watch itself'. This, however, was merely a periodical perambulation, and was put down in 1569 in favour of a 'standing watch'.

In Paris in the sixteenth century, owing to the prevalence of street robberies at night, the inhabitants, as in London, were ordered to keep lights burning before their houses, and in 1558 the police of that city ordered *fallots* (large vases filled with pitch, resin, and other combustible matter) to be set up at the street corners. Later these were superseded by lanterns, and in 1662 an Italian named Laudati obtained a concession to erect, in Paris and other French towns, booths where links or lanterns might be hired by passengers, the charge being five sous per lantern to accompany a coach and three sous per foot passenger, per quarter of an hour, the 'taximeter' being an hour-glass carried with each lantern. An improvement in the number and form of the stationary lanterns followed soon after.

In England in Jacobean times use was made of horn lanterns, both for watchmen and for street lighting (such as it was). In 1668 inhabitants of London were ordered to 'hang out candles duly to the accustomed hour', but a constant succession of threats and commands evidenced the failure of the citizens to carry out their obligations. In 1679 a strongly worded proclamation referred to the results of such neglect, since 'when night darkened the streets then wandered forth the sons of Belial, flown with insolence and wine!'.

Many continental towns at this period (including Hamburg, Amsterdam, La Hague, Copenhagen, and a number of towns in Italy) had permanent street lighting at night, Amsterdam's lights being lamps with horn lanterns.

At the close of the century London's regulations were tightened, every householder being enjoined to hang out a lamp or lantern every night as soon as it was dark, from Michaelmas to Lady Day, and to keep it burning until midnight (night-birds taking their own risk!).

In 1694 one Hemig patented a reflector which partly surrounded the open flame of the lamp and possessed a hole in the top to assist ventilation. He secured exclusive rights to light London for a term of ten years, undertaking to place a light before every tenth door between the hours of six and twelve p.m. from Michaelmas to Lady Day. Opposition forced a withdrawal of his licence in 1716 and a relapse to the old haphazard system of individual responsibility. About 1720 the Court of Common Council ordered that all housekeepers whose houses fronted any street, lane, or public highway should 'on every dark night, that is, every night from the second night after full moon to the seventh after the new moon, hang out one or more lights, with sufficient cotton wicks, to continue burning from 6 p.m. to 1 a.m., under a penalty of one shilling'. Apparently on nights when the moon should have shown but, under our English climatic conditions, did not, the wayfarer proceeded at his own risk. In 1729 a company contracted to light London's streets, householders having the option of hanging out their own lights or paying the company to do it. In 1736 the City Corporation obtained parliamentary powers to publicly light its streets with glass lamps, which were to be kept burning from sunset to sunrise throughout the year, and nearly 5,000 were erected. Evidently, by the lengthening of the hours of lighting, the city fathers had, at last, some regard for the night reveller. Was this perhaps the result of experience after a city banquet or the waylaying of some belated alderman? Despite these provisions, the link-boys were in request to accompany those who could afford them to and from social functions and places of amusement. Houses in Bath and the older parts of central London show relics of eighteenth-century lighting in the form of link extinguishers and brackets for holding oil lamps, the latter usually over the front gate. These lamps were either in the form of the more or less square street lantern or else hemispherical bowls with flat lids after the style of the railway carriage oil lamps of the late nineteenth century.

Most of the leading European centres had some public lighting provision in the eighteenth century. Dr. Martin Lister, who visited Paris in the reign of Louis XIV, speaks of lamps hung 20 feet high, 20 paces apart, in the streets, these lamps being let down for attention by a pulley arrangement. According to his description, however, these were not oil lamps but provided with candles of four to the pound, which lasted until about midnight.

Anybody who broke the glass lanterns was liable to be sent to the galleys. At the close of the eighteenth century the street lighting of Paris was associated with even more violent methods of retribution, as the revolutionary mob found the arrangement of the rope and pulley only too handy as a means of disposing of aristocrats until the substitution of the guillotine. Before that, in 1765, Paris had introduced oil lamps with reflector plates.

At Vienna, in 1780, the inhabitants had to take or send street lamps to the lamp office every morning to have them filled with oil, and had then to place them in front of their houses during the night—the lights being supplied by the Government. Later, a corps of uniformed lamplighters under military discipline was formed. In Berlin at first the inhabitants of every third house had to hang out, in turn, lanterns before their doors. Afterwards the city erected lamp posts, the lamps for which were kept lit at the expense of the inhabitants.

The first street lit by gas in London was Pall Mall (1807); Westminster Bridge was lit in 1813, and the following year the Westminster Parish set an example in substituting gas lamps for oil throughout their area. The adoption of public lighting by gas in London was followed by a rapid spread of the method; Wigan, associated with early experiments in gas lighting, followed suit in 1832.

All this did not ensue without opposition. Opponents of street lighting (and by gas in particular) in Cologne, in 1816, argued that people who did not want light ought not to be compelled to pay for it, that the horses would be frightened, that thieves would be emboldened (a rather strange argument when the whole idea of municipalities in getting the streets lit in the seventeenth and eighteenth centuries had been the prevention of robberies), and that the constant illumination of streets by night would lessen the charm of festive lighting on special occasions! The last is perhaps the only logical argument, whatever it may be worth, since the twentieth century is evidence of the decay of enjoyment through satiety.

Gas lighting by the ordinary gas jet (and oil lamps in such villages as were lit and on country railway stations) existed not very long ago, and in the closing years of the nineteenth century the lamplighter, with his ladder or his 'torch', was a familiar figure. Incandescent gas lighting was first tried in London in 1895, and high-pressure gas lighting with inverted incandescent burners arrived in 1908.

The year 1877 witnessed the real start of public lighting by electricity, when Mr. Bush illuminated Cleveland's public square with his arc lamps, which speedily 'caught on' elsewhere. Taunton claims to have been the first town publicly lit by electricity in England. Soon afterwards the battle was between the gas lamp, now equipped with the incandescent mantle, and the electric lamp—principally the arc: first the open arc, then the enclosed arc in 1893, and later the flame arc. In recent years the advance in development of the incandescent filament lamp has intensified the electric attack, the front line of which is now the gaseous discharge lamp. What comes next? Luminous curbs?

EPILOGUE

THE SYMBOLISM OF LIGHT

(Originally printed as an article in the *Toc H Journal*, November 1937.)

THE long winter nights meant a drear struggle for mere existence to primitive man; the warming rays of spring sunshine instilled a new spirit into him and everything around him—it was the giver of life. Is it any wonder that it became divinity itself to him?

Fire and light were the attributes of that divinity. Fire, when the god was in his beneficent mood, warmed man, helped him, and kept the wild beasts at bay. When the god was angry and his thunder-clouds lowered, it would kill and destroy. Is it so unreasonable that fire and light were offshoots of divinity to him? Or, grant him even a modicum of reason, when he discovered how to produce fire for himself here, if not a phase of the divinity itself, was most certainly a gift of the gods. In this he had more insight than those to whom fire and light are mere slaves of mankind.

Such an outlook, combined with the comparative difficulty of producing fire at will in early days, led, by natural sequences of thought and necessity, to the reverence with which fire was treated and the care which was taken to keep it in being. Such a reverence provides one adequate reason for placing it under the care of the priesthood. The other potent reason is the position of the priesthood, in all primitive societies, as the guardian of the arts of peace and the regulation of the community, whether under the aegis of a hierarchical government or masked under the laws and taboos of organized religion.

Just as the domestic hearth was, and symbolically still is, the focus point of the family, so the communal fire was the focus point of the tribe or community. The pioneer sons of ancient races fared forth with a portion of the tribal fire as a sacred trust and planted it in their new colony, just as an explorer of to-day plants his national standard on new lands. If the religious element is not so marked in the latter case, the symbolism is much the same.

The fire was both heat and light at first. Light, as a separate amenity, was the child of a fire—a fire in miniature. Nothing, then, is more natural than that the separate light should carry with it some at least of the divine nature of the fire. Added to this, there is the lore of the poet and the philosopher. Light is like life itself—mysterious in origin, gradually strengthening, then burning steadily or else badgered in the winds of adversity, finally diminishing, fluttering, dying away as mysteriously as it came. In itself a divine gift, it becomes, in turn, emblem of divinity, life, and the soul itself.

With so many points of contact with life, material and spiritual, it would be more surprising to find light left out of religious symbolism than to find it, as it is, present in the rites and ceremonies of nearly every form of belief on the face of the earth. Candles on the altar of a Christian church have their counterparts in the temples of ancient Egypt; time candles, suggestive of Alfred with his hour-marked candles, burn before the Bird of Dawn in China.

It is difficult to trace the use of sacred lights in any church to a single origin—the practice is so widespread and ancient that it might have been borrowed directly from a number of sources. It is difficult, too, at times, to draw the line between worship and symbolism—or even between ceremonial use and necessity. Too often the fanatic of one religion accuses the adherent of another of idolatry or fire-worship where the idol or the flame is but a symbol such as he employs himself. Often, on the other hand, what is set up by the thinker or the dreamer as a symbol becomes God himself to the ignorant. That is the danger of symbolism which has led to the extreme view which sees in it nothing but paganism and the negation of true religion. Yet symbolism, to many, has its place, just as picture writing began the exchange of unspoken thoughts.

If the followers of Zoroaster lighted candles in honour of the sun-god, it is not to be assumed that they were worshipping the flame itself. To them, the sun was the visible evidence of god and the flames of the candles were earthly symbols of the sun beyond their mortal reach. In Egypt, during the reign of the monotheist Akhenaten, lamps were lit in adoration of the Sun (Aten). At one of the Inca feasts messengers of the Sun passed through the town carrying lighted torches, to rid the place of disease, the torches being finally thrown into the river.

At a time when rejoicing and thankfulness more frequently went hand in hand, the divine element was constantly represented by light. Thus Herodotus records a feast of lights around the houses in Egypt. He says: 'At Sais, when the assembly takes place for the sacrifices, there is one night on which the inhabitants burn a multitude of lights in the open air around their houses. They use lamps in the shape of flat saucers, filled with a mixture of oil and salt, on the top of which the wick floats. These burn the whole night and give the festival the name of the Feast of "Lamps".'

Almost the same process takes place in India at 'Divali', the festival of light, dedicated to Lakshmi, the goddess of wealth. The analogy is the more striking in that the Indian earthenware lamps used to illuminate the houses of the poorer devotees are almost exact replicas of certain open saucer (spouted) lamps found in Egypt. The illumination is supposed to bring good fortune from the goddess; its neglect the reverse. Presents are exchanged and gambling forms an important part of the festival.

Another famous light festival is, of course, the Chinese Feast of Lanterns. A striking feature in China is that a debtor who has not cleared himself has a lantern borne behind him to indicate that he has 'not seen the light'. (In the western world, in these days of credit and instalment systems, the debtor often has his own headlights!) Across the China Sea, in Japan, the birth of the present heir to the throne was celebrated by a display of some ten million lanterns in processions throughout the land.

The Walpurgis Night festivities of Germany, in which torches light the scenes of revelry, are relics of pagan May celebrations associated with witchcraft but still survive; so also does the Hallowe'en (31 October) fire, substituted by candles in Ireland and flaming torches in parts of Scotland. The birthday cake, with the candle for each year of age, affords another example of the modern use of festival lights.

Such illuminations as those at Blackpool and on occasions of national rejoicing in modern civilizations may have no conscious connexion with ancient light festivals but so close is the parallel that it is difficult to be absolutely certain that there is not some traditional suggestion in the choice of such a method of celebration or the appeal which it has to the imagination. At the least, it is evidence that human thought or imagery works on traditional lines.

In Egypt the Sais festival was not the only use of lamps for religious celebrations. There are paintings of Egyptian royalty offering sacrifices in which a vessel is shown with one or more flames (apparently proceeding from floating wicks) either placed on the pile of offerings or else held in the hand of the celebrant. This may have served as an incense burner too; in fact, it is sometimes shown being fed with incense, but it is certainly a light as well. Both in Egypt and in ancient Greece oracles were consulted after the lighting of lamps upon an altar.

The seven-branched candlestick (or group of lamps) upon the ancient Jewish altar would be no novelty but a recognized feature of religious worship. To-day it has its descendant in the 'Chanukkah' lamp with its eight wicks (plus, usually, an extra pilot light) though tradition ascribes the 'Chanukkah' (dedication) itself to the time when Judas Maccabeus having won freedom and the Temple for his nation, the oil, apparently enough for one day only, miraculously burned for the week's span. To that light upon the Hebrew altar the Church of to-day may prefer to trace its sanctuary lamp and its altar lights but both are much more likely to have come from Egypt, the scene of so much of the Church's early history, including the birth of the monastic system, and the source of much ecclesiastical tradition. The glass sanctuary lamp now in use bears ample trace of its descent via Byzantium, and is a constant reminder that the age of Constantine introduced the definitely ritual use of lights into the Christian Church. To Constantine also may be indirectly ascribed the use of candles on the Christmas tree, seeing that he ordered public illuminations to be carried out on Christmas Eve. St. Lucia, among Christian saints, appropriately enough, has her own particular festival of lighted candles in Scandinavia. At first the Christian Church was opposed to the use of lights for ceremonial purposes: indeed, a fourth-century writer, ridiculing pagan rites, says: 'They kindle lights as though to one in darkness. Can he be thought sane who offers the light of candles and lamps to the Author and Giver of all light?'

The medieval church had a special rite for blessing candles at Candlemas in which the analogy of physical and spiritual fire was stressed. During Holy Week thirteen lighted candles were placed before the altar and, as the penitential songs were sung, they were extinguished one by one. Possibly this has some connexion with the Atonement. The ceremony of excommunication included a rite in which twelve priests stood about the bishop, holding lighted torches which, at the conclusion of the anathema, they cast down and trampled underfoot to symbolize the withdrawal of the light of Christian fellowship from the condemned; conversely, reception into the Christian Church was accompanied by lighted candles, just as a Toc H member to-day is received into fellowship with the Lamp of Maintenance in his hand.

Ritual lamps are used profusely in Lamaist ceremonies in Tibet and lighted lamps mark holy places in the Mohammedan world.

Apart from these special festivals of light, the use of lights is common as part and parcel of the ordinary ceremonies of communal and domestic life. When the ancient Greek married, the new family hearth-fire was kindled by torches carried by mother and mother-in-law, bearing flames from the respective parental hearths. To-day, in the Greek Church, bride and bridegroom hold candles. Is this a relic of the ancient custom? Torches figured in Persian, Turkish, and Japanese weddings; they appeared in various wedding ceremonies on the continent of Europe and possibly in old English weddings as well. The Indian wife used lights not only to bring fortune but also to gauge it. Lamps fed with coco-nut oil (usually, it seems, made of the nut itself) were floated downstream; if they disappeared, burning steadily, in the distance, all would be well, but if they failed or were wrecked misfortune was to follow.

Something of the same idea as the kindling of the new hearth is apparent not only in the carrying of a portion of the communal fire to a new colony but also in the flame at the Olympic Games of 1936, which was taken from the altar fire on Olympus and brought across Europe to Berlin by torch, passing from hand to hand of relays of runners, to light the Olympic fire which burned throughout the sports proceedings. Similarly, Armistice Day was celebrated in 1936 in Paris by the bringing to the tomb of the Unknown Soldier of a torch lit at the tomb of the Unknown Soldier in Brussels; at the same time, to the latter itself were brought flaring torches from all parts of Belgium and from the Congo.

The torch accompaniments of funeral processions may have originated in symbolism, but it is equally probable that they had some original connexion with cremation and had their practical purpose in ignition of the funeral pyre, to which the symbolism was added later. A well-known archaeologist (Mr. H. E. Balch, F.S.A., of Wells, Somerset) has, in fact, suggested that the so-called 'incense cups' (small perforated pottery vessels) of early Bronze Age burials were really tinder holders, which were swung to fan the tinder into flame for the purpose of cremation and afterwards left in the burial chamber.

If so, there is an obvious line here, whether by ancestry or coincidence, with the swung censers of temple and church, though Egyptian wall paintings suggest an Egyptian connexion for these also.

The use of torches or tapers in Christian funeral processions is sometimes ascribed to the reign of Constantine, when so much of the early simplicity of the Church was discarded, but whether this be so or not, it was certainly not an original idea, as both Greeks and Romans before had used it habitually; and the numerous finds of earthenware lamps in the Christian catacombs, many of them with slight evidence of use, if any, suggest a symbolic use beyond the mere necessities of illumination at that period, though it is more than probable that the whole idea originated with the necessity of lighting a procession after dark. Throughout the Middle Ages and in the Catholic countries to-day lighted tapers are an essential accompaniment of funeral and lying-in-state. Medieval France forecast the undying flame on the tomb of the Unknown Soldier with its 'Lanterns of the Dead' (especially prevalent in Aquitaine)—stone towers with open-work top stories in which a lamp was kept burning to invoke divine protection for the dead in God's acre and to symbolize prayers made for them by the living. Bequests made in the medieval age to provide for candles burning in chantry chapels had their counterpart in property left by Romans to provide for lighted lamps with incense at the tomb on appointed dates, and the idea may still be seen perpetuated in Italian cemeteries, where a light is kept burning in the chapels of family tombs; the lanterns in Florentine cemeteries are often replicas of the larger Renaissance lanterns or torch-holders of the palaces in the city.

No Irish 'wake', of course, would be properly carried out without its accompanying candles—properly as many as the number of candlesticks which can be begged or borrowed, but always an odd number. (Presumably a dealer in these utensils would be provided with an unusual number!) In the north of England it seems to have been customary, three centuries ago at any rate, to burn candles over the corpse even in the daytime—hence 'corpse candles'. Setting a candle at the head of the corpse is also an old Jewish custom. The use of corpse candles, in general, is suggested as again symbolic of life—the hope of eternal life; it was in this light (!) that the Early Church saw them and excused the retention of what was then admittedly a pagan practice.

In India prayer and light are also linked; lamps of small oil capacity are lit to burn during the offering of prayer only and from the trifling amount of oil some of them carried one imagines some of the prayers must have been short or hurried! Tibetan ceremonial would be incomplete without an array of butter lamps.

What are we to make of all this? Is it just a question of superstition derived from the supposed divinity of the flame—a fire-worship, in short, whether carried out in a Christian country or the little-changing East? That may be the verdict of the materialist or the extreme anti-ritualist. To the idealist and to those who would fain understand practices which are not their own, there are other aspects. Fire and light are mysterious, control them as we will. They are among the most potent gifts of Providence; without them little of our modern civilization would be possible. They are almost inseparable from human existence and intimately domesticated. Like the wind and life itself, we know not whence the flame comes and whither it goeth—it is the perfect symbol of life, the soul, and human progress; knowledge, man's greatest distinction, has kept pace with the light—first of all the crude smoky torch or the tiny struggling flicker, then the coarser brighter flares of lamp and candelabrum, in a world of ill-digested and ill-applied learning, which served but to accentuate the surrounding gloom. Lastly we reach the bright modern lighting and modern science, hand in hand, both symbolically and practically—both wide in range but apt sometimes to produce a blinding glare which makes us halt to wonder whether the softer lights of quiet thought have not their occasional uses for retreat and peace.

And so ends the history of a very human phase of man's activities—for the present.

BIBLIOGRAPHY

Collection of Heating and Lighting Utensils in the United States National Museum (W. Hough, Smithsonian Institution).
Ancient Stone Implements of Great Britain (Evans).
Life in Ancient Britain (Ault).
Journal of the Scottish Societies of Antiquaries (paper by J. R. Atkinson).
Man before Metals (Joly).
Nineveh and Babylon (Layard).
Reports on the Ur Excavations, 1933–4 (Sir L. Woolley).
The Bible.
The Illustrated London News, various issues.
Parthian Pottery from Seleucia (N. C. Debevoise).
Journal of Egyptian Archaeology, April 1924, Nov. 1931.
The Rock Tombs of El Amarna.
Ancient Egyptian Materials and Industries (Lucas).
The History of Herodotus.
Technical Arts of the Ancients (Neuburger).
Home Life of the Ancient Greeks (Zimmern).
Corinth: Terracotta Lamps (Broneer).
Catalogue of Greek and Roman Lamps in the British Museum (Walters).
Guide to Greek and Roman Life (British Museum).
Life of the Greeks and Romans (Guhl and Kohner).
Ancient Pottery (Birch).
Report of the Research Committee of the Society of Antiquaries on Uriconium.
Greek and Roman Bronzes (Lamb).
Roman Ehnasya (Petrie).
Quarterly of the Dept. of Antiquities in Palestine, vol. v, nos. 1 and 2.
Catalogue of Early Christian Antiquities in the British Museum.
Saints and their Symbols.
Oriental Customs (Burden).
Catalogue of the Arab Museum, Cairo.
Observations on Popular Antiquities (Brand).
Discovery, Nov. 1937.
The Luttrell Psalter.
More Medieval Byways (Salzman).
Parish Life in Medieval England (Gasquet).
Churchwardens' Accounts (Cox).
English Church Fittings (Cox).
Waxcraft (Cowan).
The Arts in the Middle Ages (Paul Lacroix).
The Ancient Monuments of Durham Cathedral.
English Monastic Life (Gasquet).
Guildhall (London) Museum Catalogue.
Cluny (Paris) Museum Catalogue.

BIBLIOGRAPHY

Medieval London (HOME).
The Guilds and Companies of England (UNWIN).
Ecclesiastical Art in Germany (DR. LUBKE).
The History of Southampton (DAVIES).
The Connoisseur, vol. x, no. 37.
'Domestic Candlesticks from the 14th to the 18th century' (A. D. CURLE, in *Proceedings of the Society of Antiquaries of Scotland*).
The Reliquary and Illustrated Archaeologist (various papers).
Old England.
The Legend of Montrose (SCOTT).
The Natural History of Selborne (GILBERT WHITE).
Sussex County Magazine, June 1931.
Welsh Folkcrafts (National Museum of Wales).
Iron and Brass Implements of the English House (LINDSAY).
Proceedings of the Guernsey Society of Natural Science, 1908 (paper by J. C. HOCART).
Colonial Lighting (HAYWARD).
'Pan Lamps of Pennsylvania' (DR. RUSHFORD in *Hobbies*, Sept. 1937).
'Lamps of Colonial America' (WATKINS in *Antiques*, Oct. 1937).
Lys og Lysstel (WALLEM).
Turning Night into Day (ILIN).
Artificial Light (LUCKEISH).
A History of Wonderful Inventions (TIMBS, 1870).
HAYDN'S *Dictionary of Dates.*
History of Lighting by Gas (CHANDLER).
Electric Illumination (W. T. O'DEA, South Kensington Science Museum).
Living Lights (HOLDER).
Miners' Lighting Appliances (South Kensington Science Museum).
British Lighthouses (J. S. WRYDE).
The Encyclopædia Britannica.

ACKNOWLEDGEMENTS

THE author would like to express his appreciation for their assistance to the following (and to any others he may have inadvertently omitted):

The late Professor H. Balfour, Mr. H. E. Balch, F.S.A., Mr. L. A. Benachi (Alexandria), Herr Bosenberg (Dresden), Messrs. Julius Daniels (Boston, U.S.A.), G. E. Evans (Carmarthen) and Paul Faraday, Professor Battiscombe Gunn, Messrs. Harding (Director of Antiquities, Transjordan), Hawkes, F.S.A. (Borough Librarian, Wigan), Laidler (Port Elizabeth), Oliver Myers (Egypt Exploration Society), Noall (St. Ives, Cornwall), B. R. Parkinson, M.I.Mech.E., Iorwerth Peate, F.S.A., Pryce (British Museum), and Simmons (Editor, *Distribution*), Mrs. Todd (late Achimota Museum, Gold Coast), and Sir Leonard Woolley.

Messrs. Price's Patent Candle Co., Edison Swan Electric Co., International Gas Detectors Ltd., Radstock Coal Co., and Fields.

The Science Museum (South Kensington), the British Museum, the National Museum of Wales, Guilles Allès Museum (Guernsey), the Northern Museum of Antiquities (Stockholm), the museums at Aberdeen University, Ankara, Birmingham, Madras, Winchester, Berne, Trondheim, the Neprajzi Museum at Budapest, the Royal Ontario Museum, Toronto, the Ethnographic Museum at Zagreb, the State Historical Museum at Moscow, the Abbey Folk Museum (New Barnet), York Castle Museum, the Conseil National des Femmes Hellenes (Mme Hagimichali, Athens), and the British Commercial Gas Association.

INDEX TO COLLECTIONS, ETC., QUOTED

(Plate numbers in Roman lettering)

INDEX TO COUNTRIES

(*Plate numbers in Roman lettering*)